MUNICIPAL WASTEWATER TREATMENT TECHNOLOGY

MUNICIPAL WASTEWATER TREATMENT TECHNOLOGY

Recent Developments

U.S. Environmental Protection Agency

NOYES DATA CORPORATION
Park Ridge, New Jersey, U.S.A.

Library of Congress Catalog Card Number: 92-25244
ISBN: 0-8155-1309-7
Printed in the United States

Published in the United States of America by
Noyes Data Corporation
Mill Road, Park Ridge, New Jersey 07656

10 9 8 7 6 5 4 3 2 1

Library of Congress Cataloging-in-Publication Data

Municipal wastewater treatment technology : recent developments / U.S.
Environmental Protection Agency.
p. cm.
Includes summaries of speakers' presentations from the 1991
Municipal Wastewater Technology Forum, sponsored by EPA's Office of
Wastewater Enforcement and Compliance (OWEC).
Includes bibliographical references and index.
ISBN 0-8155-1309-7
1. Sewage--Purification--Congresses. 2. Sewage disposal--United
States--Congresses. I. United States. Environmental Protection
Agency. II. Municipal Wastewater Technology Forum (1991 :
Washington, D.C.) III. United States. Environmental Protection
Agency. Office of Wastewater Enforcement & Compliance.
TD745.M86 1992
628.3--dc20 92-25244
CIP

Foreword

This book presents recent developments in municipal wastewater treatment technology, based on a forum sponsored by the USEPA in 1991. The 25 presentations included cover the areas of land treatment, sand and gravel filters, operation and maintenance, biological nutrient removal, sludge, stormwater, disinfection, constructed wetlands, and municipal water use efficiency.

Several recent and/or upcoming changes in Federal regulations will affect all of those involved in wastewater technology development and transfer. These include impending sludge and stormwater regulations and the reauthorization of the Clean Water Act. Because of technology transfer implications, the information presented here will be beneficial to those engineers, managers, and coordinators involved in treating municipal wastewaters.

The appendices to the book contain a listing of Regional and State wastewater coordinators' addresses, and a summary of innovative and alternative technology projects by state.

The information in the book is from *Proceedings of the U.S. EPA Municipal Wastewater Treatment Technology Forum—1991* (June 5–7, 1991, Portland, Oregon), prepared for the U.S. Environmental Protection Agency, issued September 1991.

The table of contents is organized in such a way as to serve as a subject index and provides easy access to the information contained in the book.

Advanced composition and production methods developed by Noyes Data Corporation are employed to bring this durably bound book to you in a minimum of time. Special techniques are used to close the gap between "manuscript" and "completed book." In order to keep the price of the book to a reasonable level, it has been partially reproduced by photo-offset directly from the original report and the cost saving passed on to the reader. Due to this method of publishing, certain portions of the book may be less legible than desired.

ACKNOWLEDGMENTS

This document was prepared by Eastern Research Group, Inc., Arlington, Massachusetts. Denise Short was the Project Manager. Technical direction was provided by Wendy Bell of EPA's Office of Wastewater Enforcement and Compliance. The text was based on attendance at the Municipal Wastewater Treatment Technology Forum, transcriptions of the presentations, and submissions made by the speakers. The time and contributions of all the Forum speakers are gratefully acknowledged.

NOTICE

Contents and Subject Index

STORMWATER

DISINFECTION

MUNICIPAL WATER USE EFFICIENCY

APPENDICES

INTRODUCTION

Update on EPA's Sludge Policy and New Sewage Sludge Regulations

Robert Bastian

U.S. Environmental Protection Agency
Washington, DC

The initial round of comprehensive sewage sludge technical regulations required by Section 405 of the Clean Water Act (CWA) were published on February 6, 1989, in the Federal Register (Vol. 54, No. 23: 5746-5902) as "Proposed Standards for the Disposal of Sewage Sludge" for public comment. The new proposed technical regulations (to be issued as 40 CFR Part 503) cover the final use and disposal of sewage sludge when incinerated, applied to the land, distributed and marketed, placed in sludge-only landfills (monofills), or on surface disposal sites. Co-landfilling of sewage sludge with municipal solid waste will be covered under the new 40 CFR Part 258 Municipal Solid Waste Landfill regulations (proposed on August 30, 1988 [53 FR 3314] and expected to be issued in final form early in 1991). Ocean dumping is to be phased out by the end of 1991 under the provisions of the Ocean Dumping Ban Act of 1988 (PL 100-68) signed into law on November 18, 1988.

The proposed Part 503 rule contains standards for each end use and disposal method consisting of limits for 28 pollutants in the form of sludge concentration limits or pollutant loading limits, as well as management practices and other requirements such as treatment works management controls over users and contractors, and monitoring, record keeping, and reporting requirements. As proposed, the requirements would apply to the final use and disposal of sludges produced by both publicly owned treatment works (POTWs), and privately owned treatment works that treat domestic wastewater and septage, but would not apply to sludges produced by privately owned industrial facilities that treat domestic sewage along with industrial waste.

Over 650 parties submitted more than 5,500 comments identifying some 250 issues in response to the Proposed Part 503 regulations. Formal comments on the proposed regulations were received from 30 states and four environmental groups, as well as many POTWs, consultants, equipment vendors, etc. During the 180-day comment period provided on the proposal (which ended August 7, 1989), experts from both inside and outside EPA were involved in thoroughly reviewing the technical basis of the proposal. The review involved experts from the Agency's Science Advisory Board, environmental groups, academia, and various scientific bodies with expertise in areas covered by the proposed rule. The majority of commenters indicated that the proposed rules were overly stringent, used unrealistic conservative assumptions, and at a minimum, will discourage beneficial use of sludge. Others raised questions about how to better define the sludge use and disposal categories, terms such as *de minimus* and "clean" sludge, and which models, risk assessment methodologies, and data to use for determining the proposed numeric limitations.

The Agency also conducted a National Sewage Sludge Survey (NSSS) to obtain better information on current sludge quality, use, and disposal practices. The survey collected information from 479 POTWs on sludge use and disposal practices and costs, and analyzed sludges from 181 POTWs for 419 analytes—all the metals and inorganics (including pesticides, dibenzofurans, dioxins, and PCBs) for which gas chromatography/mass spectroscopy (GC/MS) standards exist. These data are being used in developing regulatory impact analysis and aggregate risk analysis of human health, environmental, and economic impacts and benefits of sludge use and disposal practices to help refine the Part 503 regulations, and to help identify which additional pollutants in sewage sludge should be regulated in the future. As a result of settlement of litigation in Oregon concerning the failure of EPA to issue the new regulations by the dates specified in the Water Quality Act Amendments of 1987, EPA will identify additional pollutants and a schedule for a second round of sewage sludge rulemaking in June 1992.

The Agency published its analysis of the new NSSS data in the November 9, 1990, Federal Register Notice (55 FR 47210) for public comment. In addition, the Notice requests comments on alternative approaches that EPA is considering for various sections of the Part 503 regulations. These approaches are based on comments received on the proposed Part 503 regulations and information received since the proposal. These include:

- Revised approaches for regulating 1) land application of septage, 2) organic pollutants in emissions from sewage sludge incinerators, 3) the application of sewage sludge to non-agricultural land, and 4) the disposal of sewage sludge on a surface disposal site

- Potential changes to the input parameters for the models used to develop pollutant limits for sewage sludge applied to agricultural land or distributed and marketed

- Alternative pollutant limits (i.e., "clean sludge concept" for sewage sludge applied to the land or distributed and marketed)

- The eligibility of a pollutant for a removal credit with respect to the use and disposal of sewage sludge

The public comment period on the notice closed on January 8, 1991. EPA is utilizing the comments received on the notice, the February 6, 1989, proposal and the recommendations of the peer review panels to craft the final rule. A number of the external scientists involved in the peer review effort continue to be involved in assisting the Agency in developing scientifically defendable pollutant limits and in addressing key technical issues raised in public comments. It is anticipated that the proposed pollutant limits and management practices included in the proposed regulations, and even some of the basic approaches for regulating sewage sludge taken in the proposal, will change significantly. Current plans calls for promulgation of the final Part 503 regulations in January 1992, as a result of the Oregon court-imposed schedule.

Meanwhile, the Part 122-124 and 501 regulations, which will require the new Part 503 technical regulations (once issued in final form) to be imposed through an NPDES or state permit (issued under an approved state program), were issued on May 2, 1989, in the Federal Register (Vol. 54, No. 83: 18716-18796) as "NPDES Sewage Sludge Permit Regulations; State Sludge Management Program Requirements." A "Sewage Sludge Interim Permitting Strategy" was issued in September 1989, describing the Agency's strategy for carrying out the new Section 405 CWA requirements to impose controls on sewage sludge use and disposal practices in NPDES permits issued to POTWs until the new Part 503 technical regulations become effective. Pursuant to the "Interim Strategy," EPA or the states may issue sludge permits as agreed to by the state/EPA agreements. POTWs should consult their NPDES authorities as to the appropriate procedures and requirements. The final version of the "POTW Sludge Sampling and Analysis Guidance Document" was issued in August 1989 to provide technical guidance on the sampling and analysis of municipal sewage sludge. "Guidance for Writing Case-by-Case Permit Requirements for Municipal Sewage Sludge" was issued in final form in December 1989.

Copies of these documents are available from the Permits Division in EPA's Office of Wastewater Enforcement and Compliance.

Finally, a number of issues concerning some programmatic and technical considerations needed to implement sewage sludge beneficial use programs on federal lands have arisen among federal agencies. To remedy this, EPA is working closely with the Office of Management and Budget, the Bureau of Land Management, the U.S. Forest Service, the U.S. Fish and Wildlife Service, the Department of Defense, the Department of Energy, TVA, FDA, and other federal agencies that generate or use/dispose of sewage sludge on federal lands to establish a unified federal policy on beneficial use of sludge and to prepare guidelines that federal agency land managers can use in determining the appropriateness of land application of sewage sludge for their facilities. A Federal Register notice containing the new "Interagency Sludge Policy on Beneficial Use of Municipal Sewage Sludge on Federal Land" that has been designed to supplement the existing 1984 EPA policy promoting beneficial use of sludge and the 1981 EPA/USDA/FDA policy and guidance document should be issued by early summer 1991.

LAND TREATMENT

Best Available Technology for Design and Siting for Land Application of Wastewater on the Rathdrum Prairie—Kootenai County, Idaho

John Sutherland

IDHW–Division of Environmental Quality
Coeur d'Alene, Idaho

Introduction. Sewage was first applied to land to dispose of it or to fertilize vegetables. The practice dates back at least to 1559. These forms of disposal increased in popularity through the middle of the 19th century. In the last half of the 19th century land disposal was largely abandoned in favor of centralized treatment and discharge to surface water. These plants are designed to reduce the amounts of suspended material and oxygen demanding substances; however, contaminants such as the nutrients nitrogen and phosphorous are not well treated and may cause problems with accelerated plant growth resulting in degraded water quality. Various types of advanced wastewater treatment can be used to substantially increase removal efficiencies.

With population growth occurring in many areas, increasing wastewater flows are nearing or exceeding the assimilative capacity of the receiving waters resulting in the need for increased treatment or alternative methods of wastewater disposal. Modern wastewater land application systems, designed and operated to protect surface water quality, are in operation throughout the United States. The extent to which the protection of ground-water quality was incorporated into the design and siting is largely unknown.

Land application of municipal wastewater offers some distinct advantages to other wastewater treatment methods. Nutrients such as nitrogen (N) and phosphorous (P) can be utilized by plants, tied up in the soil, and volatilized. Heavy metals often can be adsorbed onto soil particles, and organics can be volatilized and degraded. However, if improperly sited, designed, and operated, ground-water contamination can result. When this ground water is the only source of drinking water for 400,000 people, as in the case of the Rathdrum Prairie Aquifer, informed decisions are imperative.

The Idaho Division of Environmental Quality is currently engaged in a study, using funds provided by the U. S. Environmental Protection Agency (EPA), to determine the feasibility of applying secondary treated municipal wastewater to the Rathdrum Prairie of northern Idaho. This area lies over a portion of a sole source aquifer, which the State of Idaho Department of Health & Welfare, Division of Environmental Quality (IDHW-DEQ) has determined to be susceptible to surface land use activities.

Background. The Spokane Valley/Rathdrum Prairie Aquifer, located primarily in Kootenai County, Idaho, and Spokane County, Washington, lies in a valley filled with glacial outwash deposits created when ice dams on glacial Lake Missoula breached and in catastrophic events flooded the entire area. In Idaho the Rathdrum Prairie Aquifer covers 283 mi^2 of the total 409 mi^2 basin. It is bordered by mountainous terrain and numerous lakes which provide the majority of the recharge to the aquifer. Ground-water flow is generally from the northeast from Spirit Lake and Lake Pend Oreille to the southwest where it discharges to the Spokane and Little Spokane rivers near Spokane, Washington.

The glacial outwash soils located over the aquifer are generally excessively drained, and due to the general absence of confining layers in the soil profile little protection exists for the ground water from surface land use activities.

In 1978 the Rathdrum Prairie was designated a "sole source aquifer" by EPA and in 1980 as a "special resource water" by the IDHW-DEQ.

These designations attest to the significance of this resource to the 400,000 users on both sides of the state line. The majority of residents living on the Rathdrum Prairie and Spokane Valley utilize this resource including the cities of Coeur d'Alene, Rathdrum, Bayview, Dalton Gardens, Hauser, Post Falls, Hayden, Hayden Lake, and Athol, Idaho, as well as Spokane, Washington.

Prevention of contamination is viewed as the best possible method of managing the water quality of the Rathdrum Prairie Aquifer. This area is presently experiencing rapid population growth. Combined with the climate, hydrogeologic setting, and soil characteristics, this presents some unusual challenges.

Best Available Technology. It is important to understand several concepts that are required to protect the ground water from contamination. A distinction needs to be made between the disposal of wastewater and its treatment. In using land application as a disposal method, the wastewater is applied at rates too high for hydraulic uptake and often is not treated before leaving the root zone where further treatment is unlikely to occur naturally. In a departure from standard practice, which is to load the system at the agronomic uptake of nitrogen with little or no consideration given to hydraulic overloads, Idaho DEQ feels the best available technology is to load the system at the plant's consumptive use of water. In theory this will generate little or no leachate to contaminate ground water. Application at rates in excess of the hydraulic consumptive rate may result in contaminants such as nitrate being flushed below the root zone, making them unavailable for crop uptake, and increasing the potential for aquifer contamination. An important operational parameter of this approach is that it only allows for seasonal land application during the months when plant growth is occurring and evapotranspiration rates are in excess of precipitation.

The following criteria are important considerations in siting and designing a land application system.

Hydrogeology. An understanding of the hydrogeology of a site is essential to the design of a land application system. An understanding of the aquifer characteristics and the unsaturated zone are important in determining the pollution potential of underlying ground water. Ground-water flow direction and velocities are important for determining monitoring locations. Areas of special vulnerabilities should be identified.

The Rathdrum Prairie Aquifer is recognized as one of the most productive aquifers in the country. This aquifer has high porosities, permeabilities, and transmissivities. Ground-water velocities have been calculated in some areas in excess of 50 ft/day. Recently, a well pump test for a production well discharging 6,000 gpm was pumped with a 5 ft drawdown for 8 hours near the area of a proposed pilot study. Vulnerable areas include recharge sumps where the terminal ends of streams flow out and disappear into the prairie.

<u>Climate</u>. Climate is important for several reasons. The growing season must be determined to effectively design for application rates and seasons. Precipitation is important in determining application rates. It is also important to know when evapotranspiration rates exceed precipitation. Important to the Rathdrum Prairie are the approximately 24 inches of annual precipitation, dry summers where evapotranspiration exceeds precipitation, and growing season dates.

<u>Soils</u>. Soils have the ability to store some of the water from precipitation and land application for plant growth, reducing infiltration to ground water. Soils can also retain nutrients and potential contaminants allowing for biological or chemical degradation or utilization in plant tissues. Important soil characteristics include soil depth, slope, texture, drainage, permeability, cation exchange capacity (CEC), and organic matter content.

Due to the coarseness of the glacial outwash subsoils (Missoula Flood Deposits) overlying the Rathdrum Prairie Aquifer and their rapid permeability, the surface soil layer is extremely important. The soils tend to be excessively well drained. This, combined with the nearly level (0 to 2 percent) to moderately sloping (2 to 7 percent) topography, promotes infiltration as opposed to overland flow. The surface soil horizons are formed from a combination of loess and volcanic ash in a gravelly silt loam with moderately rapid permeability. Due to the finer texture of the silty surface soil and the presence of volcanic ash and allophone ash weathering product, the surface soil has high water holding capacity and adsorption capacity for selected nutrients and contaminants if present.

The soil pH and CEC are important in that they affect the soil's ability to store nutrients and potential contaminants. Soil pH ranges from 5.6 to 7.4. The surface soils tend to have a high CEC due to a high organic matter content which allows the soil to store positively charged ions. These characteristics provide some capabilities for retaining certain chemicals (cationic and polar); however, because of the lack of confining layers, negatively charged ions (anions) such as nitrate can move with the soil water and potentially reach the ground water.

Wastewater Characterization. A complete wastewater characterization is a necessary component of a properly designed land application system. Although many of the potentially toxic elements will receive some degree of treatment (volatilization or biodegradation of organics) or are retained in the soils (heavy metals), some may have a detrimental effect on the crops, livestock, or the ground water.

Cadmium is an example of an element that is toxic to both plants and animals and if present can cause problems. Other heavy metals such as copper, zinc, and nickel are of concern for plant toxicity. It is important to prevent phytotoxicity and food chain contamination. Testing to identify potential toxics is mandatory. In areas that have industrial discharges to the treatment plant an aggressive pretreatment program is imperative. An analysis and characterization of the wastewater from two treatment plants near the Rathdrum Prairie indicated acceptable levels for land application for the Idaho DEQ pilot study.

Crop. Crop selection is extremely important to a properly designed land application system. The crop must have the ability to consume sufficient quantities of water as well as nutrients. "Down time" from irrigating due to cropping practices must be considered in design.

Different crops have different rooting depths, as well as hydraulic and nutrient needs. IDHW-DEQ feels that for the Rathdrum Prairie a split application to alfalfa-orchard grass hay, and bluegrass for seed production is most appropriate, in that differing cropping requirements will allow for continual application of wastewater during the growing season.

Pilot Project. Initial results from a literature review and office analysis are cautiously optimistic that land application is a feasible form of wastewater treatment on the Rathdrum Prairie. A pilot land application system is currently under design for the Hayden Lake area. The system will be installed instead of a community drainfield. This system will be extensively monitored by IDHW-DEQ for the next few years. Information collected should help to determine the feasibility of using land application over the aquifer as a permanent solution to wastewater disposal in Kootenai County.

Conclusion. This study has determined that for the Rathdrum Prairie Aquifer the Best Available Technology for land application is at the hydraulic consumptive use rate of the selected crop. The design and siting of the system is determined by the climate, hydrogeology, soils, wastewater characterization, and the crop.

This technology can likely be applied to many similar situations where the application of wastewater is proposed over a vulnerable aquifer.

SAND AND GRAVEL FILTERS

Sand Filters: State of the Art

Harold L. Ball

Orenco Systems, Inc.
Roseburg, Oregon

Approximately 50 recirculating sand filters have been constructed on the West Coast since 1980. Serving facilities such as apartment complexes, recreational vehicle parks, and other small communities, they range in capacity from 5,000 gal/day to 120,000 gal/day. In most cases, wastewater from dwellings is transported to the sand filter by pressure sewer (S.T.E.P.) systems or by variable-grade small-diameter gravity systems.

Effluent discharged from a recirculating sand filter should have BOD_5 and TSS consistently less than 10 mg/L. The ammonia level should always be less than 1 mg/L and nitrate nitrogen should run between 20 and 40 mg/L.

Figure 1 is a schematic of a typical recirculating filter system. The component parts are the recirculation tank with its pumps, the sand filter followed by a splitter basin, and a dosing tank. The recirculation tank (Figure 2) receives effluent from the septic tanks in the collection system. Effluent is pumped from the recirculation tank to the filter, where it is applied to the surface by means of PVC pipes with small, closely spaced orifices. Having passed through the filter media, the effluent drains into a splitter basin (or Mickey Mouse ball valve) which returns about 80 percent of the flow to the recirculation tank and diverts the other 20 percent to the dosing chamber for final discharge. The electrical controls to operate the pumps will fit easily into an enclosure (it should be NEMA 4X) the size of a suitcase; therefore, the extra expense of a building to house the controls is unnecessary.

The volume of the recirculation tank needs to equal about one half the maximum expected daily flow. The filter is sized to have at least one square foot of surface area for each 10 gal of incoming wastewater expected daily. Figure 3 is the side view of a filter.

Operation and maintenance of recirculating sand filters is inexpensive and easy. For example, an average 50,000 gal/day recirculating sand filter should require fewer than two

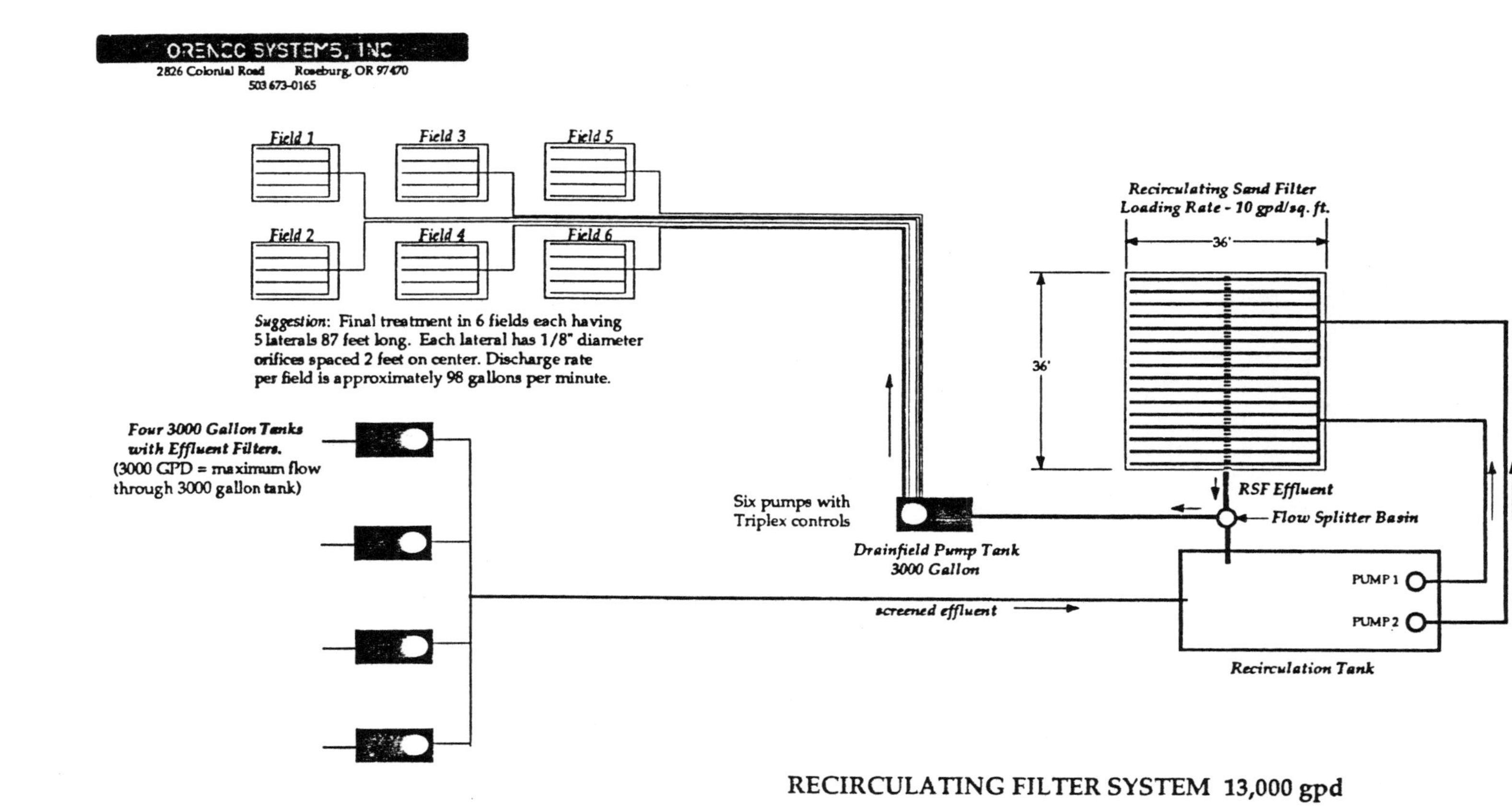

Figure 1. Recirculating Filter System

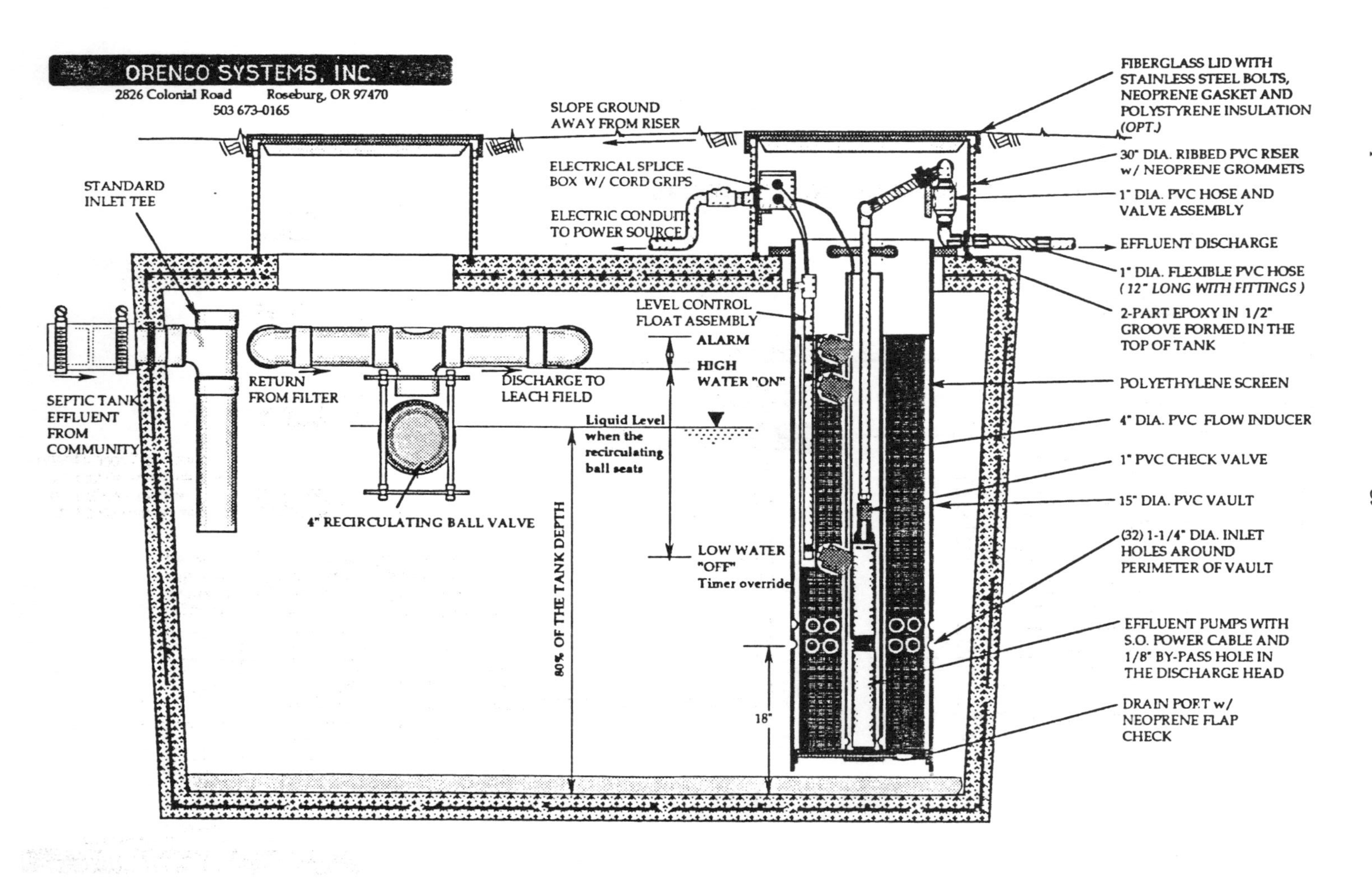

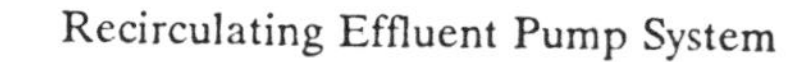
Figure 2. Recirculating Effluent Pump System

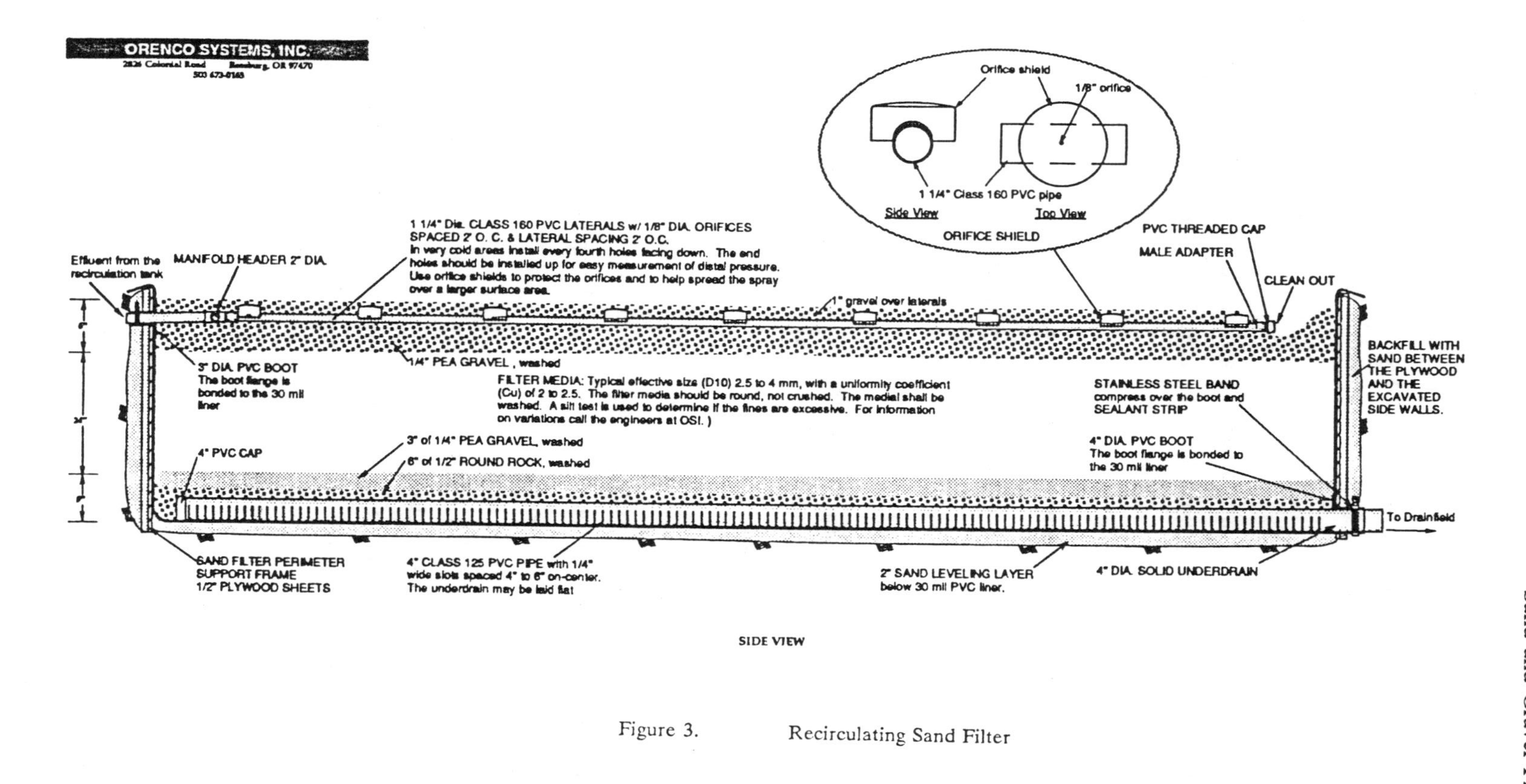

Figure 3. Recirculating Sand Filter

manhours per week. If disinfection is not required, the only other operating cost is powering and maintaining the recirculation pumps. They should be small—typically 1/2 HP—so their maintenance is relatively simple. Since all parts of the system are aerobic, odor is not a problem.

The total cost of construction of a recirculating sand filter treatment and disposal system ranges from $2 to $5 per treated gallon. Variables affecting cost are:

- Experience and skill of the designer
- Degree of cooperation from regulatory agencies
- Availability of site that is compatible with neighbors
- Topography and soil type
- Cost of sand and gravel
- Types of final disposal, e.g., drainfield, lagoon followed by irrigation, or disinfection and discharge to river or bay
- Intervention by environmental activists

Of these variables, the most critical is the experience and skill of the designer. Since a comprehensive design manual is not currently available and since the state of the art is changing rapidly, anyone contemplating the construction of a recirculating sand filter is well advised to get an experienced designer involved in the project.

University research as well as OSI's ongoing R&D program are already yielding promising results. Optimum media sizes and dosing techniques are being determined and specially designed equipment is being developed to bring down capital construction costs and make the operation and maintenance of recirculating sand filters even easier than they are now.

The Tennessee Experience with the Recirculating Sand Filter Wastewater Treatment System for Small Flows

Steve Fishel

Division of Water Pollution Control
Nashville, Tennessee

In 1987 the Tennessee Division of Water Pollution Control wrote a small flows design criteria chapter that basically required the construction of recirculating sand filters (RSF) as a primary alternative to package activated-sludge wastewater treatment facilities. At the advice of our staff of lawyers, this initiative was also written into our rules, prohibiting the construction of activated sludge plants for flows below 30,000 gpd.

We feel the benefits of the RSF alternative in performance, operability, reliability, and low O&M cost are irrefutable. The recent interest in the RSF alternative has confirmed our action. We feel one of our roles as a regulatory agency is to show leadership in promoting these benefits.

<u>Historical Small Flow (Package Plant) Problem.</u> There are 350 small (less than 75,000 qpd) domestic treatment plants in Tennessee, as compared to 250 municipal plants. These make up about 1 percent of the state's total treated sewage discharge. Because of the high number of operational complaints, and both the potential environmental and health related impacts of these discharges located in populated or high-use areas, our field offices are driven to spend a disproportionate amount of time at these facilities. The nature of these complaints involves real or potential health problems from fecal coliform flushed over the discharge weir as a result of bad operations.

The average small domestic wastewater plant in Tennessee is about 22,000 gpd. Most were built in the 1970s. Package activated-sludge plants were utilized at the vast majority of sites. Intermittent sand filters and lagoons were built in about 20 percent of the sites.

There were many compliance problems with package plants that were hard to correct. In 1986 we conducted a special O&M survey on our package plants that concluded that there was 50 percent noncompliance. Several other states confirmed similar bad performance. Advocates for package plants argued that the lack of compliance at plants was due to lack of enforcement and operator training. Limited state resources have prevented adequate enforcement and, to some degree, will always limit enforcement of these serious compliance problems.

A clear conclusion of the compliance problems of package activated-sludge plants was made only after similar experiences of visiting larger plants and making technical assistance diagnoses of municipal plant problems. The small municipal systems demonstrate a similar array of O&M problems.

In 1986 we also performed an economic analysis of the small package activated-sludge plant versus the RSF. The attached graph (Figure 4) from this analysis shows that the package plant has a hidden high O&M cost compared to the RSF. It also shows that the RSF's overall cost was much cheaper than that of the package plants! Because of the irrefutable compliance problems of the package plants, their adverse economics, and the limited enforcement resources, a decision was made to force the consideration of a more operable alternative which also had lower O&M cost—the recirculating sand filter.

<u>Evolution of the Recirculating Sand Filter.</u> In 1980, our state agency received a visit from Mike Hines of the Illinois Health Department, who originally helped invent the RSF concept. The RSF was presented as an optimized alteration of the intermittent sand filter which demonstrated better economics and performance. Mr. Hines was an advocate of reliability, operability, and economic design for small flows.

In early 1983, the consultant for Whitwell Hospital expressed an interest in a package plant alternative for their proposed wastewater discharge. Their discharge would be going into a high-use river adjacent to commercial canoeing. We suggested the RSF because an intermittent sand filter alternative was too large for the site. We visited with Murl Teske in Illinois to verify the capabilities of the RSF. The Illinois compliance data looked exemplary.

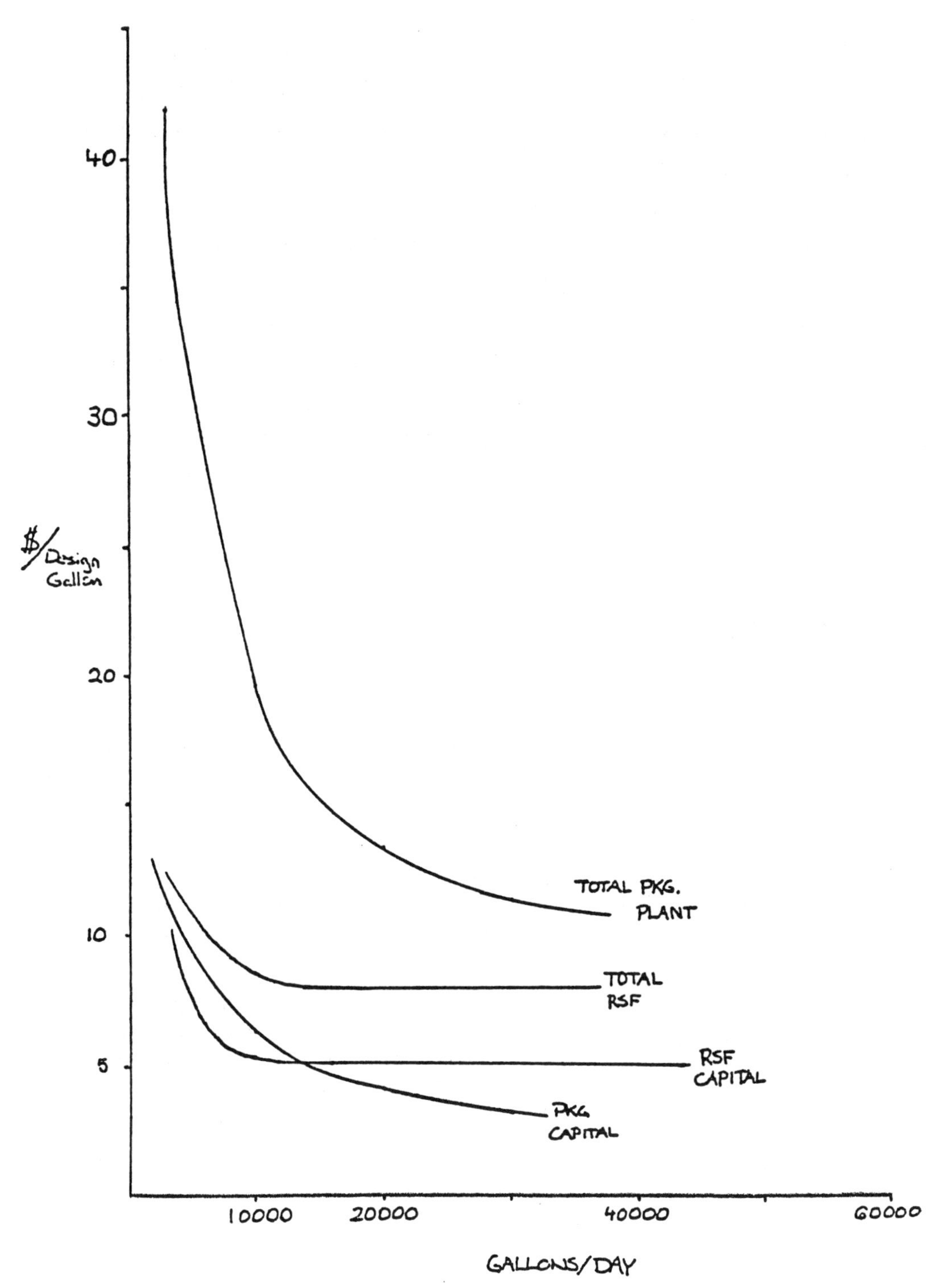

Figure 4. Small Package Activated-Sludge Plant versus RSF—Economic Analysis

The next year the Whitwell Hospital RSF was completed with the help of Mr. Hines and Mr. Teske. It was also one of the first RSFs utilizing an efficient North Carolina-style above-ground, low pressure pipe distribution system. While it was an exemplary design, it was also economically built at $6/design gallon.

The subsequent performance of Whitwell Hospital has been spectacular (see Figure 5). During close monitoring by a reputable operator, our agency, and many visitors, the discharge has been described by most as "looking like drinking water!" Nevertheless, despite the high compliance, performance, reliability, and relatively low cost, there were few advocates of this alternative in Tennessee (outside of our agency and one consulting firm). This all changed with our implementation of Tennessee's package plant alternative regulations. The attached inventory of facilities (Figure 6) illustrates the recent trend toward RSF construction. At this time, the RSF is the leading small flow alternative in Tennessee.

<u>Other Agency Actions to Improve Package Plant Alternative Designs</u>. In addition to the aforementioned items of economic analysis, small flow design criteria, and design rules, the following actions were taken:

- Our design criteria procedures were changed to emphasize the engineering report and design selection. The alternative selection was to be discussed earlier in the process. This was intended to prevent the premature submittal of undesirable alternatives as final plans.

- A new Monthly Operating Report (MOR) was designed, which closed many loopholes in reporting. This MOR requests the use of a sludge judge (a clear plastic sludge core sampler) at package plants as well as sludge disposal documentation.

- Our Division has recently encouraged the use of Oregon's RSF design concepts. Contact has mainly been via the Oregon RSF Design Criteria and Orenco Systems, Inc. The interest in Oregon designs is driven by their ability to meet ammonia standards and their use of a larger, more available gravel, as well as by their demonstrated overall performance and economy.

- The operator certification rules were changed so that sand filter operators are now designated Biological Natural System operators rather than being grouped

WHITWELL HOSPITAL RECIRCULATING SAND FILTER

	EFFLUENT				EFFLUENT		
1984	FLOW gpm	BOD mg/l	SUSP. SOLIDS mg/l	1986	FLOW gpm	BOD mg/l	SUSP. SOLIDS mg/l
JAN 84		7.5	-	JAN 86	3.0	4.0	4.0
FEB 84		5.5	-	FEB 86	7.0	4.0	3.0
MAR 84		4.0	1.2	MAR 86	3.9	5.0	7.0
APR 84		2.4	4.1	APR 86	3.6	5.0	5.0
MAY 84		1.0	3.6	MAY 86	3.4	3.0	6.0
JUNE 84	2.9	3.5	8.5	JUNE 86	3.0	1.0	5.0
JULY 84	4.3	1.8	7.2	JULY 86	3.6	1.0	1.0
AUG 84	4.2	3.6	4.6	AUG 86	3.6	2.0	2.0
SEPT 84	3.1	6.9	3.4	SEPT 86	4.9	1.0	2.0
OCT 84	4.2	4.2	7.3	OCT 86	5.8	1.0	1.0
NOV 84	3.7	4.8	7.3	NOV 86	6.8	1.0	2.0
DEC 84	3.3	8.5	5.5	DEC 86	10.2	1.0	2.0
1985				1987			
JAN 85	3.8	23.1	10.9	JAN 87	6.1	1.0	4.0
FEB 85	5.9	14.1	3.7	FEB 87	11.3	1.0	4.0
MAR 85	4.3	14.8	9.2	MAR 87	8.3	1.0	4.0
APR 85	3.9	12.3	6.0	APR 87	5.8	2.0	6.0
MAY 85	3.7	17.0	3.3	MAY 87	2.5	1.0	5.0
JUNE 85	3.6	9.8	4.8	JUNE 87	3.1	2.5	6.7
JULY 85	3.6	4.8	5.4	JULY 87	3.4	4.1	7.8
AUG 85	4.4	0.7	2.1	AUG 87	3.4	3.8	5.5
SEPT 85	3.7	0.4	2.0	SEPT 87	4.1	3.4	5.5
OCT 85	3.2	0.2	1.9	OCT 87	4.1	1.7	3.8
NOV 85	2.8	2.0	3.2	NOV 87	4.9	4.3	9.6
DEC 85	3.3	-	-	DEC 87	5.0	3.4	7.8
1988				1988			
JAN 88	11.4	2.7	11.6	JULY 88	10.1	2.2	6.2
FEB 88	5.6	3.0	7.5	AUG 88	9.2	3.2	3.0
MAR 88	4.7	3.8	3.3	SEPT 88	12.4	2.6	3.3
APR 88	4.9	2.7	6.3	OCT 88	7.7	2.4	5.8
MAY 88	5.1	2.4	4.5	NOV 88	10.8	3.5	5.0
JUNE 88	7.5	3.4	5.2	DEC 88	10.9	4.3	2.7

Most samples average from twice/month grabs. The flow data was monthly totalized average measured continuously at effluent pump station.

Jan. '85 - May '85 upset due to operational problems that were resolved. At that time the grease trap was finalized, the septic tank disposed of in May and recycle correctly adjusted.

Figure 5. Whitwell Hospital Recirculating Sand Filter

Recirculating Sand Filter Wastewater Treatment System
Facilities Located in Tennessee

Facility	Flow gpd	Loading gal/ft^2	Startup	Sand Size	Comments
Whitwell Hospital	10,000	3.0	Jan. 84	1.0 mm	Illinois design
Piney Campground	30,000	3.0	June 86	1.0 mm?	
Burns School	16,000	5.0	Sept. 88	1.0 mm.	Similar to Whitwell ERC - 333-0630
Lauderdale Inn	3,000	2.0	Sept. 89	3/8 in. gravel	Rick Dedman 901-837-3901
Westside (Macon Co.) Elementary	4,200	4.0	June 90	?	Ricky White (666-2385)
Bethesda School	17,000	3.0	Sept. 90	2.5 to 3.5	
Trinity School	13,000	3.9	Sept. 90	2.5 to 3.5	
Spencer High	9,600	3.87	Dec. 90	1 to 3 mm	
Deer Creek	20,000	5.0	May 91	Ash	Oregon & Bottom Ash
Woodgate Motel	30,000	5.0	May 91	Ash	Oregon Type & Bottom Ash
River Landing	21,600	5.0	July 91	1 to 3 mm	
Boonshill/Petersburg School	9,600	2.67	Sept. 91	3.7 mm	Linda Hinchey 370-8500
Dibrell School	4,200	~4.0	Sept. 91	not resolved	Jim Hailey 883-4933
Fairview School	15,000	5.0	Sept. 91	Ash	Oregon & Bottom Ash
Robbins School	10,000	5.0	Sept. 91	Ash	Oregon & Bottom Ash

SF/E5261151/D4/WPC-1

Figure 6. Facilities Inventory

with package plant operators. This benefits operators who do not have to learn all the activated sludge concepts.

- A program to track gravel and sand sources has been implemented. We hope to promote the use of cheaper sand such as sized bottom ash from area steam plants.

- The Division has continued to improve its information on RSF and Natural Systems design even with minimal personnel and budget commitments. This has been possible due to contributions of information from those Oregon and Illinois sources already mentioned, as well as the Small Flows Clearinghouse in West Virginia, EPA-Cincinnati, and others.

Conclusions. Small flows designs require special consideration compared to large flow alternatives. They have different economic factors and operational considerations. A comparison of reliability, operability, performance level, and overall economy between the RSF and package activated-sludge has encouraged Tennessee to promote RSF and other natural systems.

The Tennessee Division of Water Pollution Control's advocacy of natural systems by way of Division Rules and staff education has stimulated a change to a healthy diversity of alternatives such as RSF, constructed wetlands, lagoons, and spray irrigation. These alternatives are cheaper overall and are performing better and more reliably than the mechanical plants they are replacing.

Recirculating Gravel Filters in Oregon

Jim Van Domelen

Oregon Department of Environmental Quality
Portland, Oregon

Introduction. This presentation focuses on presenting the results of studies of the performance of recirculating gravel filters (RGFs) from four plants operating in Oregon. Figure 7 is a schematic of an RGF, and Figure 8 shows a cross section of an RGF. There are many variables to consider in the design and operation of RGFs:

- Media size and gradation
- Media depth
- Recirculation ratio
- Dosing method
- Loading rate: hydraulic and organic
- Expected final effluent

There is some research that is still needed to optimize design parameters versus performance. Each of these variables must be considered to determine limits, safety factors, and the performance results with a change in any one or all of the variables.

Monitoring Requirements. There are variable monitoring requirements in permits that are issued in Oregon. Some requirements include:

- System flow
- Influent BOD_5, at least quarterly during the first year
- Effluent nitrogen, NH_3-N and NO_3 (maybe TKN)
- Effluent BOD_5 and TSS

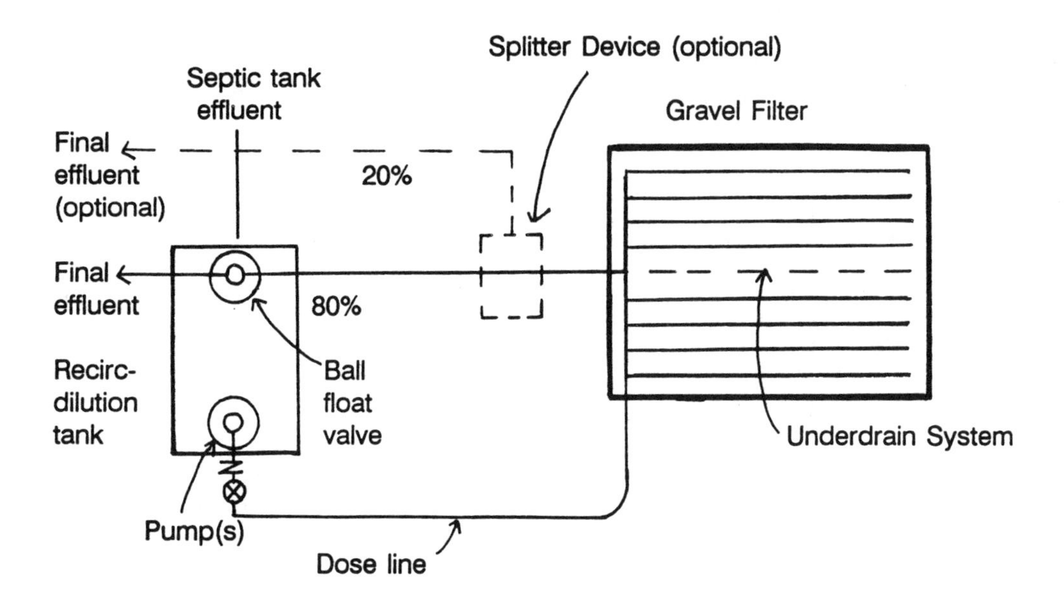

Figure 7. Schematic of a Recirculating Gravel Filter

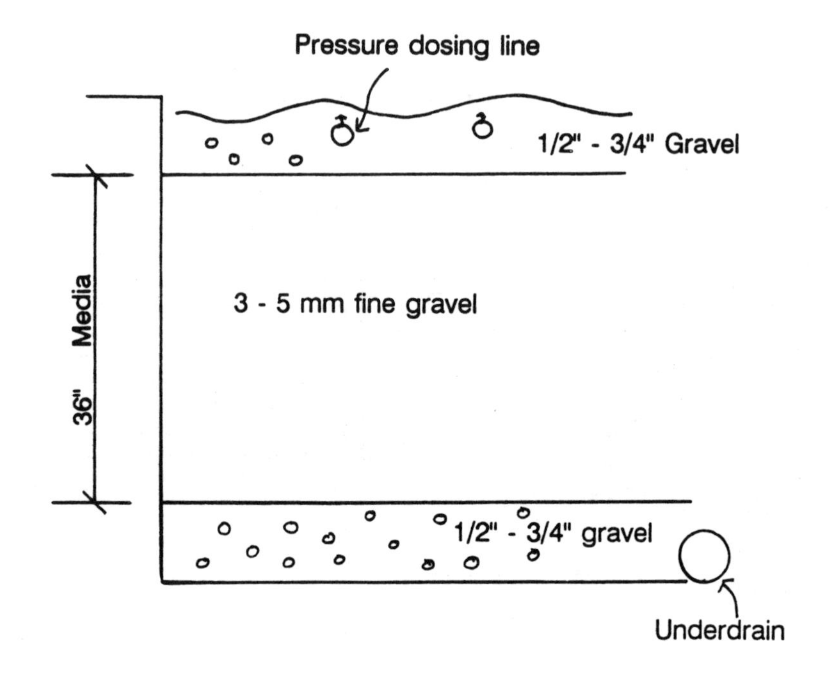

Figure 8. Cross-Section of an RGF

Design Criteria. Design criteria that should be considered include hydraulic loading (5 gallons per square foot), organics (150 mg/L, septic tank effluent), and depth of fine gravel (3 feet). Here is an example of a design criteria taken from Oregon's requirements:

- The gravel filter treatment media shall pass a 3/8-in. sieve. Less than 1 percent shall pass a No. 10 sieve. The effective size shall be between 3 and 5 millimeters. The uniformity coefficient shall be equal to or less than two (2). The material shall be a very well washed river gravel. This gravel filter treatment media must have DEQ concurrence prior to shipment to the site. The engineer shall provide us a 5 pound sample together with a gradation analysis and a particle distribution curve plotted from this data on semi-log paper. Sieves to be included in this analysis shall be 3/8 in., 1/4 in., and Nos. 4, 6, 8, 10, 50, and 100. There must be a quality assurance plan during construction for this material.

Other design criteria to consider include the recirculation ratio, the doses per day, the orifice pattern for dosing piping, the minimal burying of dosing piping, and the recirculation tank volume in gallons relative to the gpd design flow.

Operation and Maintenance. Operation and maintenance of recirculating gravel filters is an important part of achieving high performance levels. Some particular methods that have been successful include:

- Keep all equipment operating as originally installed
- Remove vegetation on filter three times per year
- Flush dosing piping and orifices two times per year
- Pump the dregs (sludge) from the recirculation tank each second year

Examples from Oregon. Table 1 presents the numbers and kinds of systems in Oregon. Table 2 presents the performance experience from four plants that are currently operating in Oregon.

Conclusion. There are many common pitfalls and traps that can be avoided by attention to the details of systematic operation and maintenance of the recirculating filters. The following list presents common traps that, if avoided, will keep the recirculating filters operating at a consistently high level of performance:

Table 1. Recirculating Gravel Filters in Oregon—May 1991

Oregon Coast

Tillamook County

Hebo Service District of Tillamook County - 1986
21,800 gpd / 4356 sq ft
Community / discharge to creek

Oregon Dept. of Corrections, South Fork Work Camp
12,500 gpd / 2500 sq ft
Forest Camp for Honor Inmates / on-site

Wi-Ne-Ma Church Camp - 1987
14,365 gpd / 2873 sq ft
Group retreat and summer camp / seepage beds

Lincoln

Oregon Parks, Beachside State Park - 1981
4500 gpd / 900 sq ft
Campground / irrigation

Whaler's Rest at Lost Creek (South of Newport) - 1984
9000 gpd / 2800 sq ft
Traveler and tourist (RV Park) / on-site

Benton County

Alsea County Service District of Benton County - 1986
30,000 gpd / 6340 sq ft
community / on-site

Coos County

Hauser Trailer Village - 1987
6800 gpd / 1296 sq ft
Mobile home park / on-site

Hilltop Restaurant (Doug Fennell) - 1985
1100 gpd / 975 sq ft (600 mg/l)
Restaurant / on-site

Table 1. (cont)

Curry County

Roque Landing (Virginia Himar) - 1986
5250 gpd / 1846 sq ft
Rest., lounge, Motel, Apts., RV spaces / seepage pit

Sandpiper Subdivision - 1985
12,000 gpd / 2500 sq ft
43 lot subdivision / on-site

Whaleshead Beach Campground (Robert L. McNeely) - 1987
8,000 gpd / 1800 sq ft
Traveler and tourist (RV Park) / irrigation

Willamette Valley

Clackamas County

Athney Creek Middle School of West Linn School Dist. # 3 - 1991
6500 gpd / 1600 sq ft
elem school / on-site

Scouter Mtn, Boy Scouts (SE 147 th & Sunnyside Rd.) - 1991
9000 gpd / 1800 sq ft
Restrooms, showers, dining facility / on-site

Damascus Dairy Queen (Jim McDonald) - 1985
1600 gpd / 726 sq ft (450 mg/l)
fast food / on-site

Fischer's Forest Park (Clackamas Sunty Service District) - 1984
10,400 gpd / 3000 sq ft
26 lot rural subdivision / on-site

Forest Park Mobile Home Park (Jim Lawrence) - 1983
5000 gpd AIRR system
Rental spaces for mobile homes / discharge to Willamette River

Orchard Crest Care Center (near Sandy) - 1989
3725 gpd / 1225 sq ft
Nursing Home / on-site

Riverside RV Resort and Spa (David A. Van Doozer) - 1987
4870 gpd / 1078 (102 Spaces)
Traveler and tourist / on-site

Salvation Army, Camp Kuratli at Trestle Glen (Barton) - 1984
8575 gpd / 2450 sq ft
Church camp / on-site

Table 1. (cont)

Lane County

Dexter Sanitary District - 1983
62,000 gpd / 12,483 sq ft
Community / on-site

Emporium, Eugene - 1988
12,200 gpd / 3025 sq ft
Warehouse and Office / on-site

Mapleton Commercial Area Owner Association - 1989
24,000 gpd / 4900 Sq Ft
Business Core of Community / Discharge to Siuslaw River

Oregon Dept. of Transportation, Oak Grove SRA - 1990
(North of Springfield)
20,000 gpd / 4032 sq ft
Traveler and tourist / on-site

Linn County

City of Mill City - 1991-92
92,500 gpd / 36,864 sq ft
Community / on-site

Polk County

City of Falls City - 1985
38,720 gpd / 7744 sq ft
Community / on-site

Yamhill County

Cove Orchard Service District of Yamhill County - 1986
11,300 gpd / 2280 sq ft
Community / on-site

Ewing Young Elementary School - 1980
3750 gpd
on-site

Mulkey RV Park (Wally A. Brosamle, Jr.) - 1990
6300 gpd / 1680 sq ft
Traveler and tourist / on-site

Table 1. (cont)

Columbia River

Clatsop County

Logger Restaurant (Richard J. Oja) - 1987
2000 gpd / 1200 sq ft (600 mg/l)
Restaurant / on-site

Riverwood Mobile Home Park (Magar E. Magar) - 1983
13,000 gpd / 4333 sq ft
Rental spaces for mobile homes / discharge to Columbia River

Westport Service District of Clatsop County - 1987
50,000 gpd / 10,000 sq ft
Community / discharge to Columbia river

Eastern and Central Oregon

Deschutes County

Redmond School District, Terrebonne School - 1989
6750 gpd / 1408 sq ft
Elem School (450 students) / on/site

Shaniko Restaurant & Hotel (Jean Farrell) - 1988
4000 gpd / 1575 sq ft
Traveler and tourist / on-site

Southern Oregon

Douglas County

City of Elkton - 1990
28,600 gpd / 6400 sq ft
Community / on-site

Oregon Dept. of Transportation, South Umpqua SRA - 1988
(North of Myrtle Creek)
2089 gpd / 520 sq ft
Traveler and Tourist / on-site

Table 2. Performance Data from Four Oregon RGFs

Dexter - 0.062 mgd (design)

Date mo-da-yr	mgd ave. mo.	BOD in/out	TSS in/out	NH3-N in/out	NO3-N in/out	TKN in/out
		----------- all effluent only data -----------				
01-25-90	.058	10	3	9.5	16	12
02-19-90	.052	5	1	5.4	12	6.6
03-12-90	.034	5	4	4.6	3.9	6.8
04-27-90	.030	11	19	5.6	24	6.7
05-16-90	.040	13	34	4.5	24	7.1
06-28-90	.039	12	37	6.8	25	7.0
07-30-90	.036	10	11	6.4	26	6.8
08-16-90	.037	9	4	3	27	3.1
09-21-90	.040	5	15	2.9	17	3.8
10-29-90	.041	5	10	10	14	12
11-28-90	.060	7	17	7.3	15	10
12-12-90	.074	5	11	9.2	6.6	10.7
01-18-91	.055	16	17	5.6	18.4	8
02-20-91	.056	7	4	1.8	16.9	2.55
03-07-91	.066	5	7	.7	15.9	1.3

****

Falls City - 0.039 mgd (design)

Date mo-da-yr	mgd ave. mo.	BOD in/out	TSS in/out	NH3-N in/out	NO3-N in/out	TKN in/out
03-15-90	.17 to	165/2	36/18	19/5	1/16	22/17
04-11-90	.22 mgd	183/18	60/8	26/5	1/22	37/7
05-17-90		164/8	52/14	8/9	1/11	20/12
06-14-90		108/3	36/7	16/5	1/29	21/7
07-12-90		76/5	3/3	36/7	1/30	42/9
08-14-90		155/32	134/18	27/20	1/4	27/21
09-11-90		80/13	44/10	28/22	1/5	33/23
10-18-90		48/36	31/25	21/23	1/12	25/25
11-13-90		95/20	80/13	29/13	1/10	35/15
12-11-90		36/8	34/13	280?/6	1/12	31/6
01-16-91		40/4	5/10	17/3	1/12	21/5
02-14-91		135/2	32/3	46/2	3/16	51/2
03-14-91		195/5	51/5	24/2	1/16	26/3

Table 2. (cont)

Westport - 0.050 mgd (design)

Date mo-da-yr	mgd ave. mo.	BOD in/out	TSS in/out	NH3-N in/out	NO3-N in/out	TKN in/out
01-02-90		100/5	24/5	43/9	-/16	
02-05-90		170/5	25/4	23/-	-/-	
03-05-90		225/3	37/5	51/-	-/-	
04-02-90		230/2	36/6	-/10.6	-/4.5	
05-01-90		235/5	22/7	35/-	-/-	
06-04-90		145/18	42/9	18/11.28	-/17.75	
07-02-90		135/6	20/11	44/1	-/4	
09-05-90		45/3	22/10	39.3/5.7	-/10.3	
10-01-90		80/3	68/4	38/6.9	-/4.3	
11-01-90		30/2	68/6	5/0.15	-/12.8	
12-10-90		20/8	13/6	35/8.4	-/3.8	
01-02-91		85/2	16/3	36/9	-/5.3	
03-01-91		120/2	48/5	35/-	-/-	

****

Hebo - 0.022 mgd (design)

Date mo-da-yr	mgd ave. mo.	BOD in/out	TSS in/out	NH3-N in/out	NO3-N in/out	TKN in/out
03-06-90	.012 to	330/8	23/11			
03-21-90	.017 mgd	228/15	46/4			
04-04-90		130/3	24/1			
04-19-90		174/17	58/8			
05-02-90		156/14	60/11			
05-17-90		162/4	42/4			
06-06-90		384/6	260/6			
06-25-90		291/38	28/5			
07-10-90		273/11	70/5			
08-09-90		-/-	56/3			
08-21-90		130/4	40/20			
09-11-90		85/26	45/3			
09-25-90		60/5	49/2			
10-17-90		94/8	50/1			
10-31-90		276/3	205/1			
11-14-90		102/3	52/1			
11-29-90		99/3	41/1			
12-27-90		22/1	8/1			
01-02-91		141/10	24/8			
01-15-91		65/1	20/2			
02-05-91		384/11	100/5			
02-19-91		117/10	10/4			
03-06-91		204/48	12/6			
03-20-91		150/9	41/4			

- Ignoring the collection system and disposal system when designing the treatment system
- Incomplete explanations of RGF design criteria to designers and engineers and owners
- Not collecting influent data before design
- Underestimating BOD_5
- Not collecting effluent (monitoring) data after years of operation
- Excessive flow to the system, collection system defects
- Failing to install the standard media
- Placing *any* filter fabric within the media: Don't do it!
- No operation and maintenance (O&M) manual
- No operator
- Failing to remove weeds and plant growth
- Failing to keep dosing piping free of debris
- Irretrievably (excessively) burying of dosing piping so flushing orifices is not practical
- Tinkering with pumps and controls like timers, splitter boxes, alarms, floats, etc.

OPERATION AND MAINTENANCE

Assessment of O&M Requirements for UV Disinfection

O. Karl Scheible

HydroQual, Inc.
Mahwah, New Jersey

Ultraviolet (UV) disinfection systems are being widely considered for application to treated wastewaters, for both new plants and retrofitting existing plants in lieu of conventional chlorination facilities. The technology is relatively new, with most plants installed over the past three to four years. It has generally been successful, although there had been many problems with the systems installed in the early to mid-1980s. Subsequent "second generation" designs have resolved many of the earlier issues, resulting in a higher degree of reliability and a more rapid acceptance of the technology. These use modular, open-channel configurations in place of the fixed, closed shell arrangements typical of the earlier designs.

An assessment of the UV process has been made, focusing on the newer designs that utilize open-channel, modular configurations. It is a part of the Office of Wastewater Enforcement and Compliance Control's (OWEC) program to provide technical assistance to local governments in the area of municipal wastewater treatment by evaluating specific technologies and reporting on their capabilities and limitations. Information was compiled from the EPA, Regional, and state offices, literature, equipment manufacturers, and wastewater treatment plant personnel. The report presents an assessment of the status of the technology relative to the type and size of UV facilities that are currently operating, and discusses the trends in system design, configuration, and operations. The design and operation of selected plants are then reviewed; this information and current practices are then summarized to give a perspective of key considerations that should be incorporated into the design of UV facilities. This presentation addresses the portion of this study that focused on an evaluation of several plants.

A total of 30 plants utilizing UV disinfection were selected for a detailed assessment of their UV process design, operation, and maintenance. Only those with open-channel configurations were chosen, in keeping with the focus of this evaluation. A random selection was made, constrained by the desire to have plants of varying size, alternate system designs, and representation by several manufacturers. The information was compiled through the summer of 1990 on the basis of supplier data and direct contact with the plant owner, operator, and/or engineer. The 30 plants are identified in Table 3.

Overall, of the 30 plants selected for evaluation, all are designed to treat to nitrification levels at minimum, and several have tertiary filtration. These conditions suggest that, in general, the facilities using UV have advanced secondary or tertiary processes, yielding effluents that are especially conducive to the application of UV. Eight of the plants use the vertical lamp configuration, somewhat higher in proportion to the horizontal configuration than is apparent in the overall census. Eight of the 30 are facilities that have retrofitted their UV systems into existing chlorine contact chambers, a procedure that is becoming popular, particularly as plants are being upgraded and with larger facilities. Generally, the plants vary in their capacity relative to design. About a third each are at approximately 25 percent, 50 percent, or 100 percent of their design capacity.

<u>Description of the UV Systems at the Selected Plants</u>. The selected plants show a certain consistency in their configurations. One to three channels are used, with a single channel in the smaller plants and the multiple channels found with the larger plants. Most of the larger systems have some flexibility in operating banks of lamps within the channel, although this is not always the case. Redundancy to any degree is not typical; only 5 of the 30 plants have redundant systems, and 4 of these are with the smaller plants. Flexibility appears to be limited, with little ability to isolate a portion of the system for repair or replacement. Bypasses were not evident with most plants, suggesting a difficulty with repairing/shutting down channels when only one channel exists.

Sizing of the units appears to be relatively consistent, falling between 0.5 and 1.7 mgd/kW, with an average essentially equivalent to 1.0 mgd/kW. This is demonstrated in Figure 9, which presents the peak design flow of the plant as a function of the total UV power (kW at 253.7 nm) of the UV system. There is some scatter, but the slope of the relationship closely approximates 1.0. Thus, a rough sizing estimate can be made for a given plant by assuming 1 kW of UV output for each mgd of peak design flow. This would be for advanced secondary plants at minimum and peak to average flow ratios less than 2.5. The 1.0 kW is the nominal UV output, equivalent to approximately 37 long lamps (1.47 m or 58 in. arc length) or 74 short

Table 3. Summary of Design Sizing/Performance Characteristics for Selected Plants

Plant	Peak Design Flow (mgd)	Peak to[h] Average Ratio	% Cap	Permit[a] Limit (100mL⁻¹)	UV System: Type	UV System: Number Channels	UV System: Number Banks	UV System: Design Peak Loading (mgd/kW)	Effluent Quality[c]: Typical Level (mg/L): BOD	TSS	NH_3-N	Bacterial[d] (100mL⁻¹)
1. Waldron, AL	1.0	1.2	80	None	H	1	2	1.58	-	-	-	-
2. Bridgeville, DE	1.2	1.5	25	200	V	1	2	0.38	6	4	3.8 (TKN)	25
3. Dakota City, IA	0.5	1.0E	83	200	V	1	1	0.6	<10	13-15	-	160-180
4. Cave City, KY	1.5	2.5	50	200	H	1	2	0.88	5	3	-	<10
5. Edgewater, MD	0.5	1.0E	16	14	H	1	2	0.47	6	5	0.9 (TKN)	1.8
6. Clearsprings, MD	0.4	2.0	50	200	H	1	1	0.47	<10	<10	-	<2
7. Leadwood, MO	1.5	3.0	30	200	H	1	1	5.0	7-14	4-10	-	<1
8. Olla, LA	0.3	1.0	100	25	H	1	2	1.12	5	0.5	-	<1
9. Dewey, OK	1.1	2.75	55	200	H	1	1	0.86	<20	78	-	31
10. Stoney Creek, VA	1.5	2.5	60	200	V	1	4	1.0	4	5	-	10-40
11. Petersburg, WV	1.8	3.0	83	200	H	1	2	1.69	<10	<10	<0.1	<20
12. Ozark, AL	5.25	2.5	43	1,000	V	2	-	1.16	16	20	0.4	<100[e]
13. Jessup, MD	1.6	1.0E	70	200	H	1	2	1.65	<3	<3	<1TKN	<4
14. Lebanon, MO	3.5	1.5	91	400	H	1	2	0.82	-	-	-	In Comp[f]
15. Abbeville, LA	4.5	2.75	130	200	H	1	2	0.86	4	8	1-2	<
16. Hanover, NH	7.0	3.0	57	240 TC[b]	H	1	2	0.82	<30	<10	-	200-220[g]
17. New Providence, NJ	6.0	3.0	25	200	H	3	2	0.69	10-20	5-15	1-2	<100
18. Owasso, OK	3.0	1.25	54	200	V	2	3	1.32	10	10.1	-	<5
19. Highspire, PA	3.8	1.9	50	200	V	1	4	1.11	<10	<10	-	30-40
20. Accomoc, VA	2.9	1.25	94	200	H	1	1	0.96	<10	<15	-	<100
21. White Sulphur Springs, WV	4.0	2.5	70	200	H	1	2	0.85	5-8	3-5	<1	<10
22. Williamson, WV	5.0	1.7	33	200	H	1	1	0.93	16	11	-	45
23. Athens, AL	13.0	1.9	96	1,000	H	1	2	1.68	8	13	2.4	324
24. Gunnison, CO	6.7	1.6	24	6,000	V	3	2	0.98	<10	<10	-	<200
25. East Chicago, IL	36.0	2.4	92	200	H	2	2	1.03	1.7	5.2	-	12
26. Olathe, KS	25.0	4.0	27	200	H	2	2	3.25	12-17	6-12	0.8	<10
27. Okmulgee, OK	5.0	Average	54	200	V	2	3	0.47	<10	7-12	<1	<200
28. Willow Grove, PA	17.5	2.5	98	300	H	1	2	0.85	5-10	10	0.5	In comp[f]
29. Warminster, PA	16.0	2.0	62	200	H	1	2	0.85	5	<10	1.1	43
30. Collierville, TN	7.0	2.0	51	200	V	2	4	1.17	10	4-10	-	3.10

(a)Fecal coliform, unless otherwise noted. 30-d average is shown, unless noted otherwise. See Table C for full permit description.
(b)Total coliform; value is not to exceed, any sampling.
(c)Average levels typical of performance. Based on discussion with operators.
(d)Fecal coliform, unless otherwise noted.
(e)Levels when second stage nitrification in operation. FC levels increase to 700 to 800 FC/100 L without nitrification.
(f)Data not available. Stated to be in compliance.
(g)Total Coliform.
(h)E designates plants with equalization.

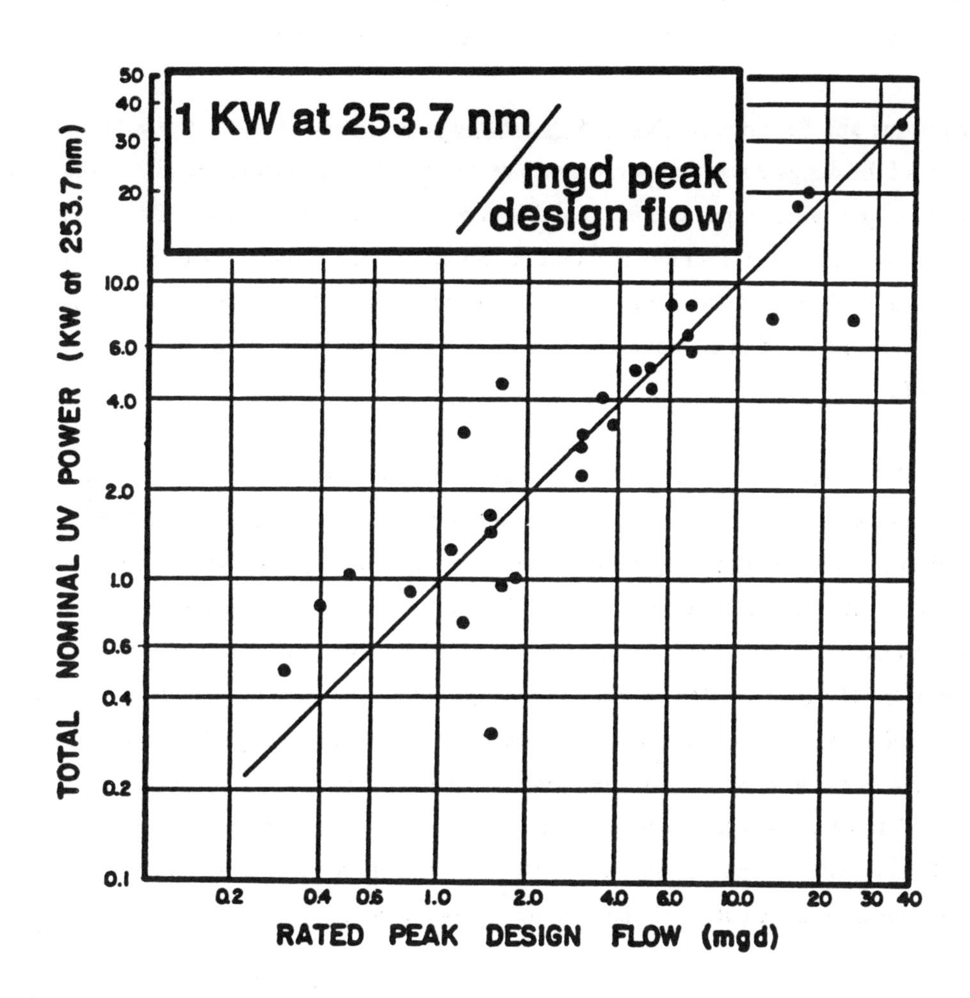

Figure 9. UV System Sizing for Selected Plants as a Function of Peak Design Flow

lamps (0.75 m or 30 in. arc length). Such an approximation should only be used in screening type assessments and should not serve as a final design sizing parameter. Note also that redundancy or standby capabilities would be added to this estimate.

Design Sizing and Performance Summary for the Selected Plants. Table 3 is presented as a summary of the design sizing and performance record for each of the selected plants. Each of the plants is generating a quality effluent and is in compliance with its permit. Those that are accomplishing a high degree of nitrification are also discharging minimal levels of coliform. In cases where the BOD and TSS levels tend to be at levels greater than 10 mg/L, the effluent coliform levels also tend to be more pronounced, with measurable densities between 10 and 200 FC/100 mL.

UV disinfection efficiency is very dependent upon the quality of the effluent generated by the upstream processes. As higher levels of treatment are accomplished, the UV process is more efficient, resulting in the need for less hardware, or providing for a greater factor of safety. Thus nitrification, denitrification, filtration and other tertiary processes that are added to conventional secondary treatment operations are particularly conducive to assuring the success of the UV process. The impact on water quality is generally represented by lower coliform densities, increased sensitivity of the bacteria to UV, and increased UV transmissibility at 253.7 nm by the wastewater. An interesting observation made from this assessment was the lack of data regarding the incoming coliform densities and the transmissibility of the effluent. The plants did not measure these parameters, even in cases where there may have been difficulties and the data could be used for troubleshooting.

Summary of O & M Practices at Selected Plants. The replacement cycle could be estimated fairly well for the lamps. It is based on the operators criteria for replacement and accounts for seasonal/year-round use of the system, and the probable system utilization rate. Thus if the system is operated on the basis of flow, the utilization would be approximately 50 percent; this would increase up to 75 to 100 percent if the system was operated manually and was basically kept in full operation as a matter of convenience or to assure compliance.

The criterion for failure is generally lamp failure and or increasing coliform densities (except at those plants with fixed operating cycles, as discussed earlier). Generally, it appears that the latter condition would be the final trigger. The high operating life cycles that are being obtained suggest that the lamps will not fail (i.e., electrode failure, shutoff); rather, their output will deteriorate to such a degree that there is insufficient germicidal energy for effective disinfection. The lamps are replaced at this point to restore the system efficiency. For design purposes, a reasonable estimate of operating life would be 14,000 hours; thus the replacement rate in a system with year-round disinfection, and an average 50 percent utilization, would be approximately 30 percent per year:

$$((8{,}760 \text{ hours/year})/(14{,}000 \text{ hours/lamp})) \times 50 \text{ percent} = 31.2 \text{ percent}$$

With the smaller systems, and to a lesser extent the larger plants, it appears that the tendency is to operate the full system (75 to 100 percent utilization) at all times instead of controlling it on the basis of flow. This would increase the replacement rate for the above example to 50 to 60 percent per year. If disinfection is required on a seasonal basis the replacement rate is reduced to 25 to 30 percent per year.

Regarding the quartz sleeves and the ballasts, it is not possible to make a direct assessment of their expected life cycle. The experience with full scale systems, particularly with respect to the open channel submerged units, is limited, covering a period of approximately five years. This is not sufficient to evaluate in-field experience for long-term replacement rates of the quartz and ballasts. Many of the replacements currently reported by operators have been due to breakage and electrical wiring failures, reasons that do not speak to the degradation or failure of the components themselves.

The quartz will degrade due to solarization of the quartz structure, resulting in a cloudiness of the quartz and a loss of transmissibility. Abrasion of the surface due to long-term exposure to the wastewater is also a contributing factor to their deterioration. There is no current feedback of replacement of the quartz for these reasons. At this point, an estimate that may be appropriate is a replacement rate of 10 years, to account for minimal breakage and for deterioration of the quartz.

Similarly, there is little experience with ballast failures and replacement rates. Earlier failures have been attributed to improper electrical design and the lack of proper ventilation in the ballast cabinets. These difficulties appear to have been corrected, although there are still reports of electrical problems with a few installations upon startup. For purposes of life cycle assessments with UV disinfection systems a 10-year replacement period is suggested.

The effort required for replacement of these key components (largely the lamps themselves) is relatively low. This is also shown in Figure 10, which presents the hours spent per year against the number of lamps that would replaced per lamp. The mean is 0.4 hours per lamp, or 24 minutes per year. There is significant variability, with the rate ranging from approximately 10 minutes to 50 minutes per lamp. Note that this is total labor, even if two people are engaged in the activity (which tends to be typical).

This analysis can be used in screening the labor and parts replacement costs for UV systems. One should be careful to acknowledge how the system will likely be operated in terms of utilization; recall that the tendency is to have much of the system on at a given time, regardless of the flow. Also account for the year-round versus seasonal disinfection requirements. Note also that these charges could be incurred in discrete intervals, rather than be spread out somewhat evenly over a period of time. This results from the likelihood that the operators will change out the lamps in total, triggered by the overall operating time and a decrease in disinfection efficiency, as discussed earlier.

A second labor factor is the time required, on a yearly basis, for activities other than replacement of the lamps/quartz/ballasts and cleaning. These would include system monitoring and sampling, area maintenance, component repair/replacement, etc. This tends to be a factor of two to six times the amount of time estimated for the replacement of key components. When added to the parts replacement activities, the total time required outside of routine cleaning needs (discussed later) is estimated. These data are plotted on Figure 11, which presents the total hours per year as a function of the system size. There is some scatter, particularly with the smaller plants. For the 14 plants with less than 150 lamps, the mean labor requirement was 120 hours per 100 lamps. The equivalent mean for plants with more than 150 lamps was 55 hours/100 lamps.

Upstream devices such as screens are used to protect the lamp battery from debris that may reach the UV system and cause damage to the quartz/lamp assemblies. Other problems

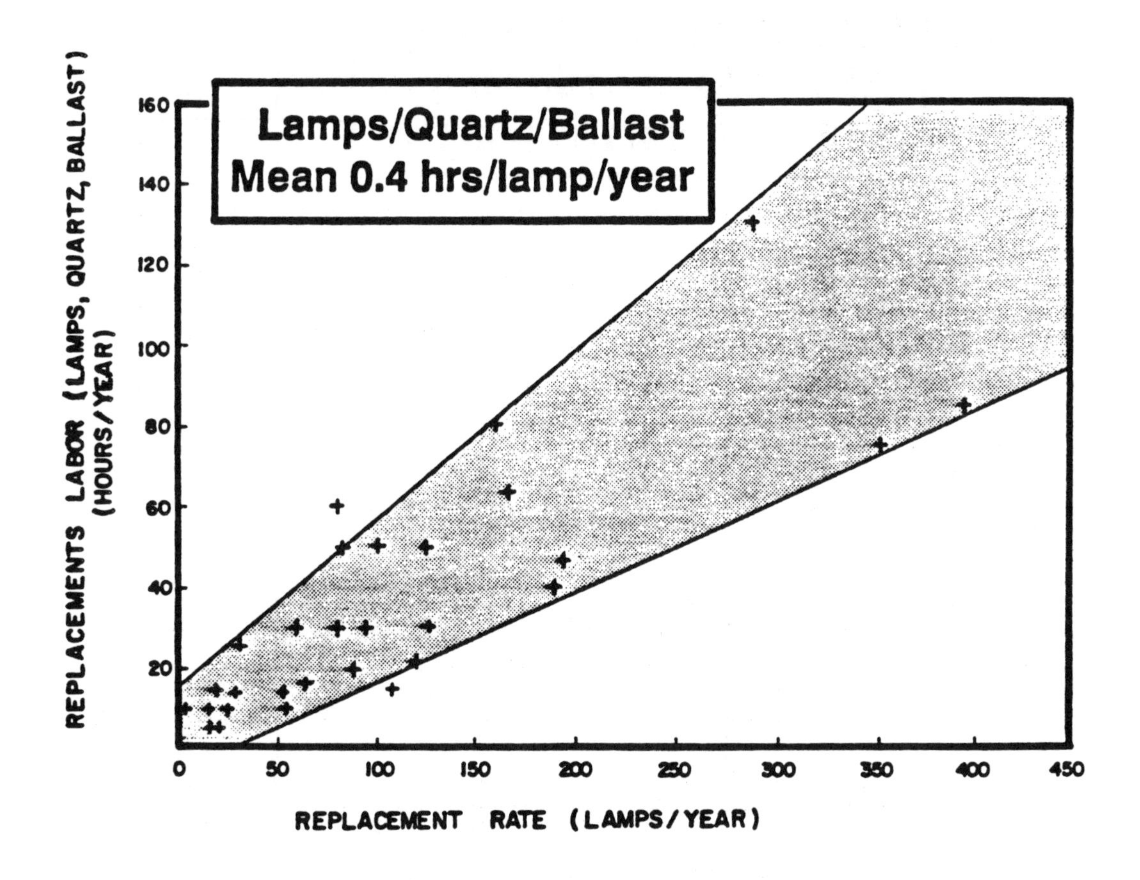

Figure 10. Labor Requirements for Replacement of Lamps/Ballasts/Quartz

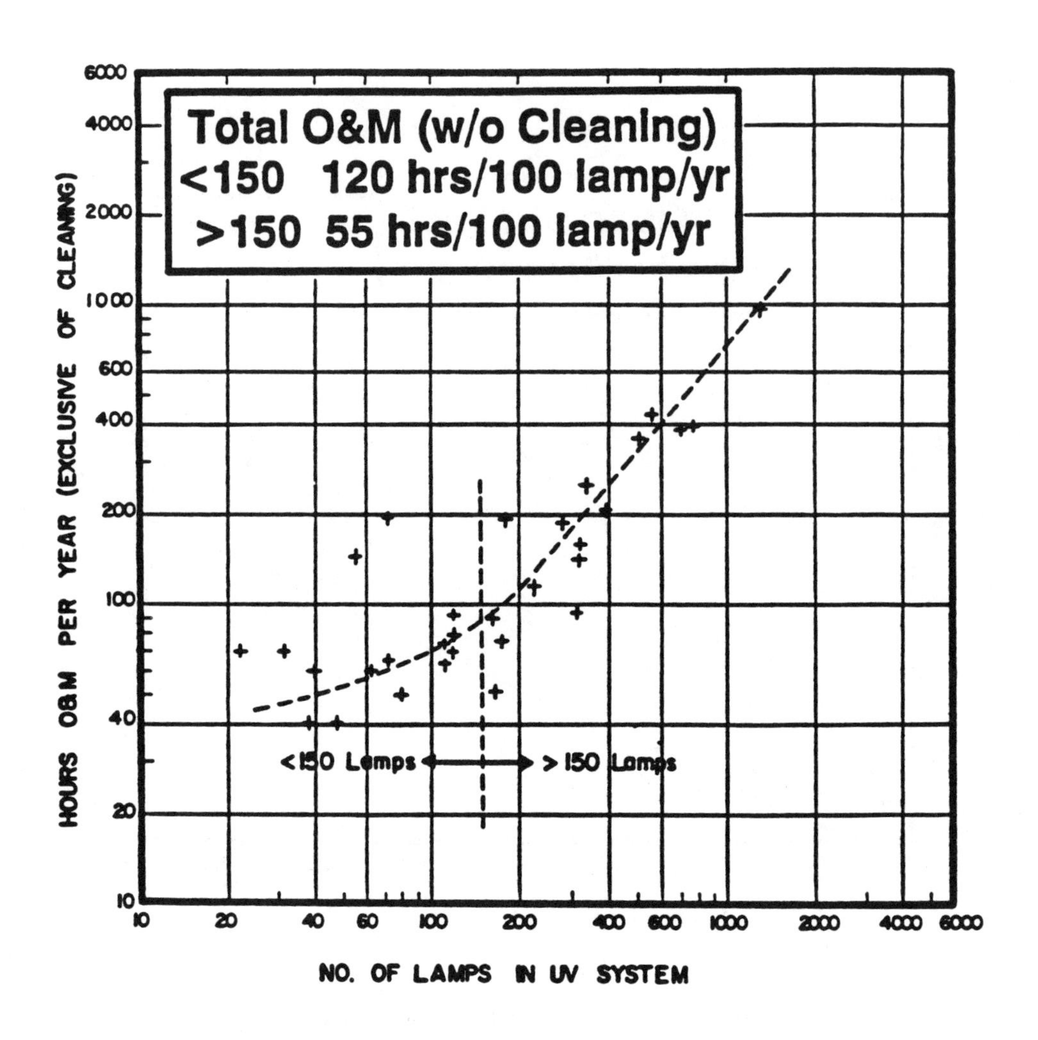

Figure 11. Estimate of O&M Labor

occur from algae sloughing off the clarifiers and leaves falling into the channels; these catch on the lamp modules and accumulate, creating additional head loss problems and maintenance tasks.

Overall it appears that the installation of an upstream screening device is an option that most plants do not choose. From this assessment, however, it also appears that it is most appropriate to have one in place. These can be simple, large-mesh (0.5-in.) screens (stainless steel), that can be slipped in and out of the channel manually for cleaning on a frequent basis. This will save considerable labor if the alternative is to clean the debris attached to the individual modules. An alternative device that may be more convenient to the operator would be a bar screen that can be raked (a moving mesh or bar screen that is self-cleaning would not be cost-effective); this would still have to be removed periodically for a thorough cleaning. Note that it is important to remember that these devices, particularly as they accumulate material, will impose a headloss; this must be accounted for when considering the hydraulic design of the facility.

A critical operating requirement is that the water level in the channel must be kept fairly constant. If it fluctuates widely (greater than plus or minus one inch from the control level), several problems can occur. In horizontal systems the top row of lamps can either be exposed or the depth of water above this row can become so great that disinfecting effectiveness of the unit is compromised. In vertical units this same problem occurs, except that the top portion of each lamp is affected.

Most plants use a mechanical level control gate to maintain the desired level. These rely on field setting and adjustment of the counter- weights to assure the proper level control over a range of flow rates. They have generally been very successful and comprise the dominant method for level control in open-channel systems. Problems are noted, however, at low flows and at plants that have no flow at times. The gates will oscillate and cause wide fluctuations in level. They are not designed to be watertight and will allow the channel to drain during periods of very low or no flow.

The method of level control should be carefully considered in the design of a facility. The mechanical gates would be the preferred device in most cases, particularly larger systems in which multichannels are used and the channel velocities can be maintained within a reasonable operating range. If there are low flow (or no flow) periods, fixed weirs may be more appropriate. Sufficient weir length must be provided, however, to avoid excessive level fluctuation. This can be accomplished by using serpentine weirs and weir launders. An alternative is to use a motorized adjustable weir slaved to a level sensor.

System control has generally been kept simple with the newer open channel UV units. This has been limited to pacing the operation of multiple channels and banks to the flow rate. The manner in which the UV system is controlled should be a function of the type and size of plant. Above all, it should be kept simple; the objective is to conserve the operating life of the lamps (and the associated power utilization). This becomes increasingly important with the larger plants (greater than 150 to 200 lamp systems), and more practical. With the small plants, it may be best to have the full system in operation, exclusive of the redundant units incorporated into the design. Manual control and flexibility should be available as the system increases in size, enabling the operator to bring portions of the system (i.e., channels and banks) into and out of operation as a function of flow and performance. Automating this activity becomes advantageous as the system becomes larger, using multiple channels.

Safety is important in the operation of UV systems, centering primarily on protection from exposure to UV radiation. This affects the eyes with a temporary condition known as conjunctivitus, or "welder's flash," that can last for several days, causing a painful burning sensation. Bare skin also will be burned upon exposure to UV at these wavelengths. Exposure risk is generally minimal, as long as the operating lamps are submerged and the lamp batteries are shielded. The danger arises if the lamps are operated in air; this should never be necessary except under extraordinary circumstances. Systems should be equipped with safety interlocks that shut off operating modules if they are removed from the channel. Electrical hazards are minimized by the inclusion of ground fault interruption circuitry with each operating module. This feature is typically standard with current systems and should be a requirement with all specified systems.

The precautions against exposure to UV radiation are straightforward. UV blocking glasses, with side shields, should be worn at all times in the general area. One plant reported that the shields were ineffective and switched to goggles for full protection. Exposure of skin should be minimized, using long sleeved shirts and buttoned necks, as examples. Signs should also be posted near the equipment and in the general area that warn of the hazard and instruct the use of glasses, at minimum. Of the selected plants, most all required and actively used eye protection, generally preferring goggles. One plant imposed stricter rules after an eye injury had occurred. Signs are also posted in several of the plants. Specific training is not typical, except that which is given by the equipment manufacturer during startup, and this is not always the case. At best, safety issues and training relating to the UV system should be incorporated into the plant's normal safety program.

Summary of UV Cleaning Practices at Selected Plants. Maintaining the quartz surfaces is a critical element in the successful operation and performance of the UV process. This is a simple task, entailing routine cleaning of the quartz sleeves with a standard agent. It is one that has at times been overlooked, however, resulting in apparent failure of the UV process because the quartz surfaces have become fouled and have lost their transmissibility. The fouling is most often due to the deposition of inorganics such as calcium or magnesium carbonates and iron. Greases or biological films can also adhere to the surface. The key task is to anticipate this and to have a fixed protocol for maintenance of the quartz surfaces.

The assessment showed considerable variability amongst the plants, making each case somewhat unique. Essentially all are successful, using methods that are relatively simple, easily applied, and which fit specifically to the conditions of the facility. This is a marked improvement from the earlier system configurations using closed shell, fixed in-channel, and teflon pipe designs. These systems suffered serious problems relating to the ability to keep the quartz or teflon surfaces clean and the access to the quartz for such maintenance tasks.

Note that the use of dip tanks is gaining favor and is generally supplied with most new systems, including those using horizontal lamp configurations. These can be in a fixed location or rolled on wheels to each bank of modules. An example is shown in Figure 12, which is a sketch of a typical unit used for horizontal lamp modules. Modules are removed individually from the channel and placed in the recirculating bath. It is then hung on the rack above the tank to drain, where it can be physically wiped and/or rinsed with clean water. In certain cases, a cage system is being devised to enable removal of banks of lamps from the channel (via a moving overhead hoist) and placement in a large dip tank. This is especially useful at larger

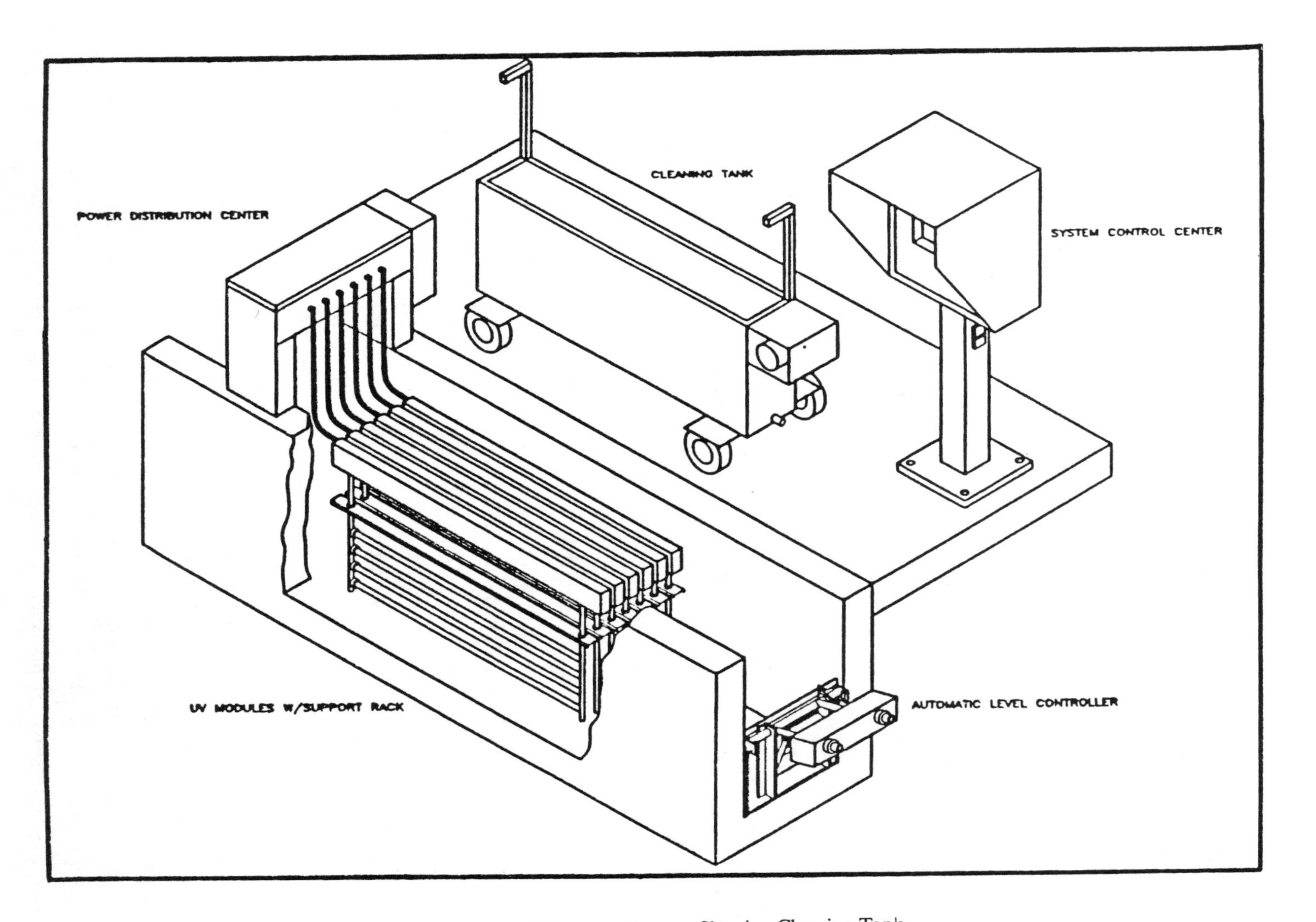

Figure 12. Schematic of UV Channel System Showing Cleaning Tank

plants. At present this is planned for the Nuese River plant in Raleigh-Durham, North Carolina, and the LOTT plant in Olympia, Washington.

Cleaning practices are highly variable. The principal points are summarized on Table 4, addressing the equipment and methods used for cleaning, the cleaning agents, the criteria used for cleaning, and the resultant frequency and labor use.

The dominant practice is to remove the modules from the channel, with or without provision of a rack to hang the module. In-place recirculation or dip tanks are more typically used for the vertical lamp module systems. The standard practice for manually cleaning the units is to simply apply the cleaner onto the quartz and then rinse the module with clean water.

Citric acid and Lime-Away are typically used as cleaning agents, although several others are used including detergents and other dilute acids. There is no hard criterion that sets the type of cleaner; the manufacturer will generally recommend one or more. It becomes a matter of trial and error specific to the plant site. This is also the case with frequency; as noted, this varies widely and depends on the specific site requirements.

The criterion for cleaning is typically based on fecal coliform densities. This was the case for two thirds of the selected plants. The remaining third was split between using the intensity meter reading, or simply setting a proscribed frequency.

In summary, the following observations are made:

- Removal of the modules is appropriate and probably best for most plants. Cages are suggested for larger plants for removing bundles of lamp modules.
- Moving hoists/cranes will facilitate removal of the module bundles or vertical lamp modules.
- Dip tanks provide a convenience and assist in cleaning modules removed from the channel.
- In-place recirculation is effective, particularly for vertical lamp modules. Agitation should be provided during the recirculation cycle. Plant should still plan to remove the modules once per year for a rigorous cleaning.

Table 4. Summary of Cleaning Practices for the 20 Selected Plants

	NUMBER OF PLANTS	COMMENTS
A. Equipment Used for Cleaning		
In-place recirculation	4	All vertical lamp modules; remove once/year
Mechanical wiper	1	One of four "in-place" units
Dip tanks	2	
Remove modules onto a rack	5*	
Remove modules	19*	No special equipment to hold the module
B. Cleaning Agents		
Citric acid	9	2 dip tanks, 4 in-place, 3 external modules
Lime-Away	10	Commercial product
Dilute HCI acid	4	
Detergent	3	Dishwashing detergent; Windex; a plant also uses Brillo pads
Phosphoric acid	2	
Sulfuric acid	1	
Tile/Bowl cleaner	1	Commercial product
C. Frequency (cycles)		
Weekly (52/year)	2	
Monthly to biweekly (12 to 26/year)	14	
Six weeks to yearly (1 to 9/year)	14	
D. Labor per cycle/per 100 lamps		
1 to 10	24	mean, 4.3 hours/cycle/100 lamps
greater than 10	6	mean, 17.4 hours/cycle/100 lamps
E. Criteria for Cleaning		
Fecal coliform	20	
Intensity meter	5	
Routine	5	

* Method is to rinse, apply cleaning agent, rinse, and return to channel.

- **The cleaning agent(s) that suits the facility is dependent upon the nature of fouling. A trial and error series of tests should be conducted, using readily available, off-the-shelf commercial products, frequency of cleaning will be dependent on the specific site requirements.**

- **Small-scale piloting would be very effective in establishing the cleaning agents and frequency most suitable to a specific plant.**

- **Monitoring fecal coliforms is a effective tool for determining the need for cleaning lamps. Note that this is also used for triggering lamp replacement.**

Frequency and Labor Requirements for Cleaning. The frequency of cleaning is highly variable, ranging from once per week to once per year. Table 4 presents the estimated time spent per year for cleaning the quartz, based on input from the operators. It is not appropriate to simply include this in the O&M labor requirement summarized in Figure 11. Rather, the time required per 100 lamps is normalized to the cycles per year.

There is no clear trend in this value relative to plant type or size. The labor requirement ranges from 0.7 to 26 hours/cycle/100 lamps. Eighty percent (24 of the 30 plants) are less than or equal to 8.3 hours/cycle/100 lamps, with a mean value of 4.3 hours/cycle/100 lamps. The remaining 6 plants range between 10.4 and 26 hours/cycle 100 lamps, with an average of 17.4 hours/cycle per 100 lamps. The overall 30 plants is 6.9 hours/cycle/100 lamps.

Overall, a value to 5 to 10 hours/cycle/100 lamps would appear to be appropriate for use in screening a facility labor requirement for cleaning. Actual yearly requirements will than depend on the frequency. Of the 30 plants, the median frequency was approximately one per month or 12 times per year. Using a median estimate of 5 hours/cycle/100 hours and 12 cycles per year, the yearly requirement would be 60 hours/100 lamps. When compared to the labor requirements on Figure 11, this is equivalent to one-half that of the larger plants and one-half that of the smaller plants. Thus, the cleaning activities can comprise one-third to one-half the total labor requirement of O&M.

Trickling Filter Operation and Maintenance Issues

Russell J. Martin

U.S. Environmental Protection Agency—Region 5
Chicago, Illinois

Within the Water Division of Region 5 is a team of engineers that maintains one of the most active and aggressive federal municipal onsite assistance programs in the nation, providing training, technical support, and consultation services for some of the 4,281 municipal wastewater treatment facilities in Region 5. Region 5 staff have learned a great deal in the last five years in this program, including successfully addressing measures that limit the wastewater treatment plant performance at trickling filter facilities. This paper will discuss our experiences at four of these facilities.

Fredericktown, Ohio.

Hardware—0.2 MGD rock media trickling filter wastewater treatment facility.

Initial evaluation—The wastewater treatment had difficulty in achieving $CBOD_5$ limits of 25 mg/L for 6 out of 12 months preceding our assistance. The clarifiers were dark and had previously been identified as anaerobic by the state agency. One trickling filter was down due to freezing problems (plant evaluated in December). A great deal of leakage was evident around the central shaft of the operating trickling filter. The WWTP recycle levels were all operating at close to maximum levels 24 hours per day. Much pump wear was evident as three pumps were down for repair.

Problem Identification—Dark color in clarifiers, pump, and trickling filter shaft was due not to low dissolved oxygen (DO), as evidenced by a multitude of positive DO readings, but to a combination of excessive recycle rates, grit, and ineffective grit removal.

The trickling filter was not operating in an effective manner due to central shaft leakage, freezing problems which halved use of trickling filters, and an inadvertent bypass in the distribution box.

Maintenance could be more effective if it wasn't spent entirely on pumps repair/ replacement.

Remedies:

1. Recycle flows reduced through the use of timers
2. Wind barriers built around the trickling filters
3. Trickling filter central shafts were retooled and new seals were installed
4. Distribution box inadvertent bypass eliminated
5. Grit removal system being installed
6. Maintenance program revised

Preliminary Results—In 1990 25 percent improvement in $CBOD_5$ effluent values was demonstrated.

Dupo, Illinois.

Hardware—0.6 mgd plastic media trickling filter wastewater treatment facility.

Initial Evaluation—The wastewater treatment plant had difficulty in achieving $CBOD_5$ limits 4 out of 12 months preceding our assistance, and suspended solids excursions also periodically occurred. Much corrosion was evident at the facility. The valves were rusted closed, steel covers in some cases were almost completely rusted through, and the structural integrity of the trickling filter was also of concern. The trickling filter and primary clarifiers were both performing poorly. One of the primary clarifiers would periodically completely surcharge the effluent weirs.

Problem Identification—Poor maintenance was the result of delayed maintenance activity. This was due to contractor problems which delayed turnover of the plant to the Village and a low Village priority for WWTP staffing. The trickling filter performance was being hurt by inadequate and inconsistent wetting. Modifications to and maintenance of nozzles was complicated by the walkways and handrails and other unnecessary hardware. The primary clarifier performance was being impacted by uneven hydraulic loading in the primary diversion box. Also, sludge was not being removed in sufficient quantities or in a consistent manner from the primary clarifiers.

Remedies:

1. Emphasized importance of proper maintenance and protection of investment with the Village fathers resulting in a greater investment of time at WWTP. A 50 percent increase in manhours at plant was obtained.
2. Installed flow restricter at one primary clarifier to equalize flows.
3. Opened the trickling filter recycle pipe.
4. Improved the solids control program by maintaining a regular schedule for solids removal.

Preliminary Results—Eighteen percent reduction in $CBOD_5$.

<u>Linton, Indiana</u>.

Hardware—0.9 mgd rock media trickling filters.

Initial Evaluation—The wastewater treatment plant was out of compliance 90 percent of the time with frequent bypassing of overloaded primary system.

Problem Identification—This plant is home to a very unusual pipe which carries wastewater in both directions, switching flow directions several times on some days. During dry weather, the pipe carries 0.5 mgd of trickling filter recycled back to the head of the plant.

During wet weather, the flow goes in the other direction through a bypass pipe where excess flows are bypassed. This recycle flow was recommended during a previous WWTP evaluation.

Remedy—Reduce the amount of trickling filter recycle.

Results—The wastewater treatment plant is now in compliance 90 percent of the time. Bypassing has been reduced by 37 percent.

Johnstown, Ohio.

Hardware—0.75 random dump plastic media two-stage trickling filter with a Dynasand final filter.

Initial Evaluation—The wastewater treatment plant was out of compliance 11 out of 12 months for $CBOD_5$ and 6 out of 12 months for NH_3. The village had replaced superintendents on three occasions in the previous 2 years.

Problem Identification—Comprehensive in-plant $CBOD_5$ tests identified that the first-stage trickling filter was not performing as expected. The failure of this first stage restricted the second-stage nitrifying ability. An examination of actual operating conditions and design parameters identified an extremely low wetting rate. Also, long in-plant holding time resulted in an increased NH_3-N value through an intermediate clarifier between the first- and second-stage trickling filters.

Remedies:

1. Repiped an existing 950 gpm intermediate clarifier pump to provide more recycle
2. Installed bigger value to increase second-stage trickling filter recycle
3. Discontinued use of intermediate clarifier for dry weather flow and used basin to capture stormwater

Results—Fifty-four percent reduction in $CBOD_5$ levels. Fifty-two percent reduction in NH_3-N levels.

Summary.

1. Take the time to evaluate carefully all trickling filter wastewater treatment plants.

2. Review trickling filter operation as an element of a wastewater treatment facility. This technology is not as much of a "stand alone" process as activated sludge.

3. Proper and consistent wetting rates are very important where plastic media is utilized.

4. The importance of proper maintenance cannot be overemphasized.

5. Recycle flows frequently are a critical element in poor performance at trickling filter facilities.

6. Proper municipal onsite assistance can result in a significant increase in trickling filter performance.

Update on the Microbial Rock Plant Filter (MRPF)

Ancil J. Jones

U.S. Environmental Protection Agency—Region 6
Dallas, Texas

This emerging and promising technology utilizing natural processes for municipal wastewater treatment is the result of research conducted by the National Aeronautics and Space Administration (NASA) at the Stennis Space Center (SSC) in Southern Mississippi over the past 20 years (1). This technology utilizes aquatic and semi-aquatic plants, microorganisms, and high surface area support media such as rocks or crushed stone. Communities, consulting engineers, state agencies, and EPA Region 6 have continued the development.

The technology was developed to treat and reclaim wastewater for reuse in space stations. On Earth, it is a low-cost, cost-effective technology for small communities, on-site treatment, individual systems, and industrial wastewater. Haughton, Benton, and Denham Springs, Louisiana, are the first applications of this technology in Region 6. Long shallow rock filters are heated by solar energy maintaining biological activity rate during cold months.

Scientific Basis. The scientific basis for municipal wastewater treatment in a vascular aquatic plant system combined with a microbial rock filter (MRF) is the cooperative growth of both the plants and microorganisms associated with the plants and rocks. A major part of the treatment process for degradation of organics is attributed to the microorganisms living on and around the plant root systems and the rock filter. Organics are held in place by the rocks and plant roots where microorganisms are given time for assimilation (see Figures 13 and 14).

This technology grows only selected plants in wastewater. The rock filters the wastewater in conjunction with the plant roots. Hydroponics is defined as "the growing of plants in a nutrient solution and without soil."

MICROBIAL ROCK PLANT FILTER

LONGITUDINAL SECTION

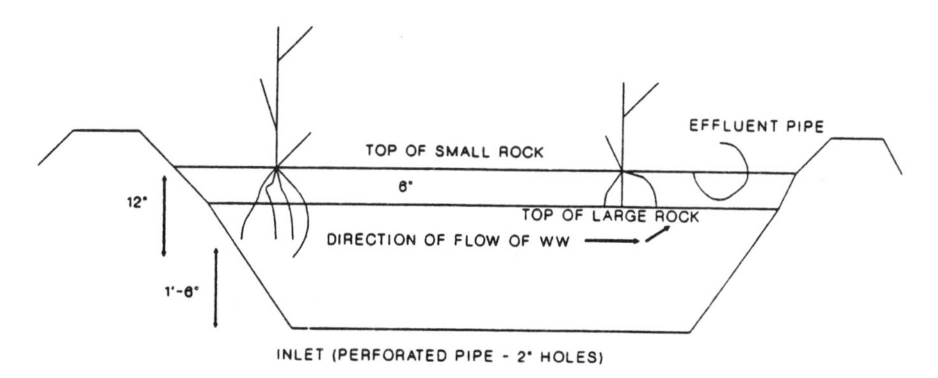

ROCKS AND ROOTS FILTER SOLIDS FROM WW AND
HOLD FOR MICROBIAL DIGESTION AND ASSIMILATION

Figure 13. MRPF—Longitudinal Section

MICROBIAL ROCK PLANT FILTER

CROSS-SECTIONAL AREA

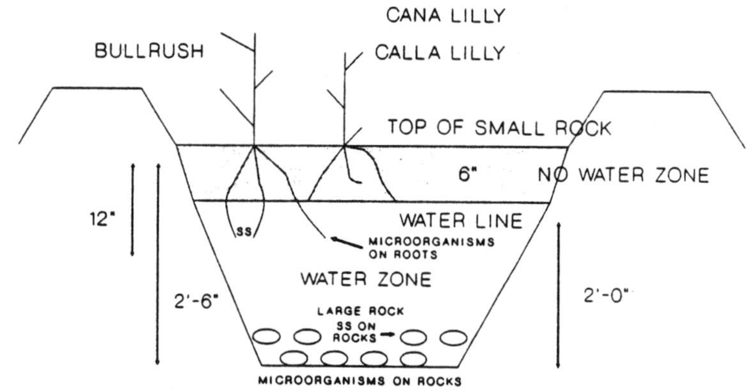

PLANT ROOTS ARE OPPOSITE CHARGES TO SS & HOLD THE SS TO ROOT
SS ARE HELD ON ROCKS BY FLOW OF WW AND GRAVITY

ROCKS AND ROOTS FILTERS SOLIDS FROM WW
AND HOLD FOR MICROBIAL DIGESTION AND ASSIMILATION

Figure 14. MRPF—Cross-Sectional area

Definition. This technology combines the application of hydroponics and the MRF technologies. The rocks (inert) support the plants and roots in a nutrient solution (wastewater). Thus the technology is appropriately defined as a microbial rock plant filter (MRPF).

It is defined by some as a subsurface flow constructed wetland. But this technology is not a derivation of the wetland technology. It is a combination of two technologies: the MRF + hydroponics = MRPF. The MRPF uses different size filter media and design philosophy than the surface and subsurface flow systems described in the EPA Design Manual "Constructed Wetlands and Aquatic Plant Systems for Municipal Wastewater Treatment" (EPA/625/1-88/022). The MRPF is a different design concept and requires a different operation and maintenance and plant management program than constructed wetlands.

Plant Functions: Aquatic Plants Translocate Oxygen. Aquatic plants have the ability to translocate oxygen from the upper leaf areas into the roots producing an aerobic zone around the roots where aerobic conditions can be maintained. Bulrush *(Scirpus Californicus)* roots in Region 6 have been measured up to 20 inches in length. Canna Lily *(Canna Flaccida)* roots have been measured up to 12 inches in length. Less oxygen is measured near the bottom of the filter. Aquatic plant roots are also capable of absorbing, concentrating, and in some cases, translocating toxic heavy metals and certain radioactive elements resulting in removal from the wastewater (See Figure 15) (2,3,4,5). Caution is to be taken to ensure proper disposition of hazardous harvested material.

Aquatic Plants Absorb Organic Molecules. In addition, aquatic plants have the ability to absorb certain organic molecules intact where these molecules are translocated and eventually metabolized by plant enzymes as demonstrated with systemic insecticides (6).

Biological reactions that take place between environmental pollutants, plants, and microorganisms are numerous and very complex, and to date, are not fully understood. But there is enough information available to demonstrate that aquatic and semi-aquatic plants serve more of a function than simply supplying a large surface area for microorganisms as some have suggested.

MICROBIAL ROCK-PLANT FILTER
PLANT/ROOT FUNCTIONS

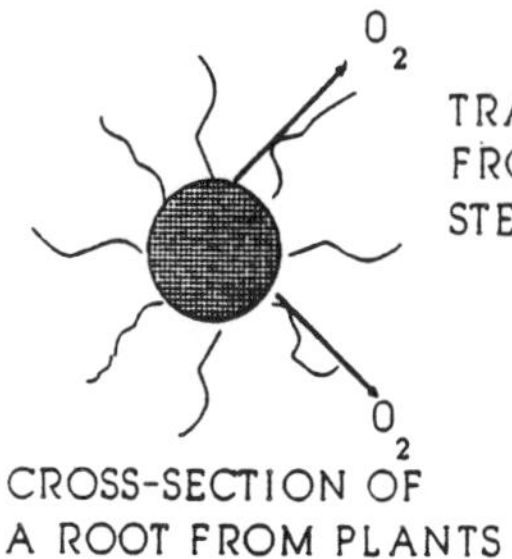

PRODUCTS OF MICROBIAL DEGRADATION OF ORGANICS ARE ABSORBED BY PLANTS.

MICROORGANISMS USE SOME OR ALL METABOLITES RELEASED THRU PLANT ROOTS

OPPOSITE CHARGES OF COLLOIDAL PARTICLES (SS) AND PLANT ROOTS CAUSE SS TO ADHERE TO ROOTS. SOLIDS FROM WASTEWATER ARE REMOVED, DIGESTED, AND ASSIMILATED BY MICROORGANISMS AND ROOTS.

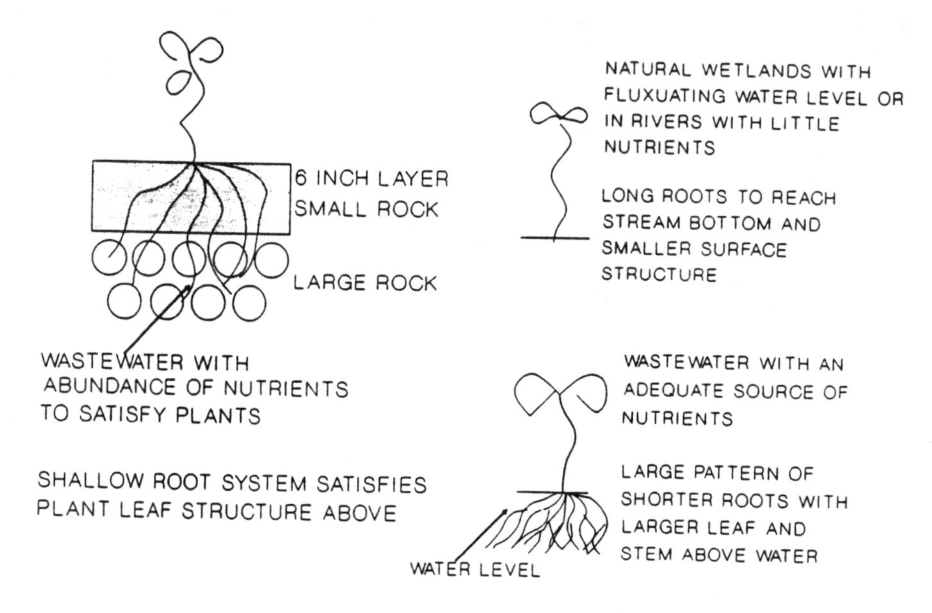

Figure 15. **MRPF—Plant/Root Functions**

Natural Regenerative Habitat. Each existing ecological system, using the other's waste products, provides for a natural regenerative habitat and sustains accelerated removal of organics from wastewater.

Removal of Suspended Solids. The vertical and horizontal flow of the wastewater holds the solids on the rocks and roots. Charges associated with plant root hairs attract the colloidal particles (suspended solids) with opposite charge. The solids are held in place for microbial digestion (see Figure 15).

Growing Plants. By definition, to benefit from this technology, plants must be grown in a nutrient solution only under conditions that maximize inter-process capability and symbiotic relationships with microorganisms and rocks. Only plants selected are to be grown in the filter. All other growth should be removed. This is not the case with the constructed wetlands philosophy.

Plant Management. Unlike the constructed wetlands, plants in the MRPF must be managed. Roots must be controlled and maintained as to depth and area to sustain the void ratio required to maintain subsurface plug flow through the filter. The plant root system grows to satisfy plant super structure (see Figure 15).

If the roots are not controlled, the filter void ratio will be reduced and cause surface flow and ponding. Roots have been measured up to 20 inches in a 30 inch filter, drastically reducing the void ratio and causing severe ponding and short circuiting. The planting density also affects the void ratio (see Figure 16).

To control root depth and area, control the plant height and umbrella. Horizontal growth varies with the umbrella of plant above the water line. Length of roots (depth below water line) varies with the height of the plant above the water line. It is suggested that the plants should be restricted to an 18-inch height above the water and the root's length and area below be measured every 6 months. The umbrella of the plant above the water should also be measured. Since terrestrial plants are utilized, this control relationship must be established (see Figure 17).

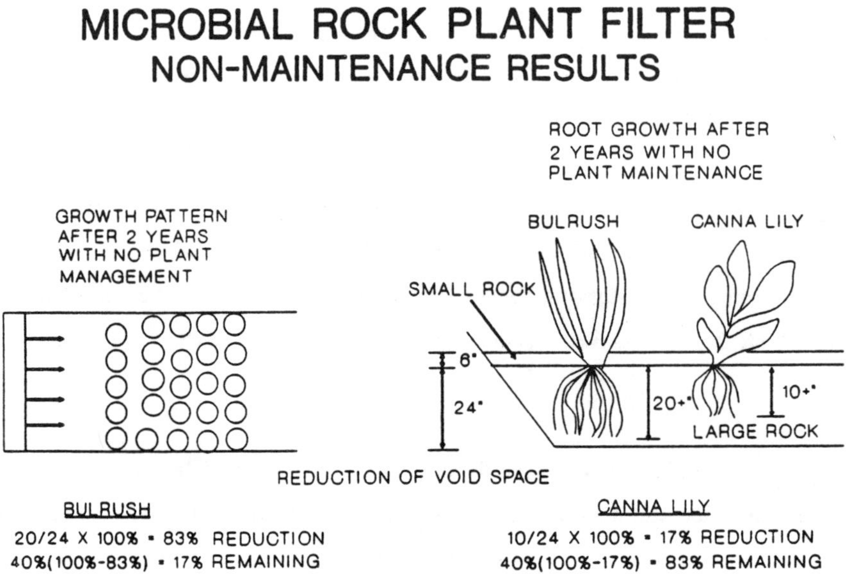

Figure 16. MRPF—Non-Maintenance Results

<u>Assign Volume of Filter to Roots</u>. Assign a volume of filter for plant roots and monitor every 6 months. Selective removal of plants may be necessary to maintain the void ratio for subsurface flow and proper function of plants and microorganisms. The planting density must also be maintained to control the proper density (see Figure 18).

<u>Design</u>. A major emphasis inherent in the design of the MRPF is the greater attention placed on multi-objective planning, inter-media impact considerations, and total systems design. Satisfaction of these objectives requires a higher level of multi-discipline in the systematic screening and evaluation of alternatives than has been generally employed in the past. More important, however, is the much greater effort needed in concept development and the formation of management alternatives. Where performance data are lacking, a pilot scale might be necessary to define operational parameters.

<u>Design Considerations</u>. A major factor in design is that there are 59 different elements of design to be considered in this complex system, of which failure to consider one or more may alter the symbiotic relationship of the natural processes which may change the results of the synergistic effect, and which may result in a risk with unacceptable consequences.

Design elements to consider are not limited to the ones in the following list, but the list will provide a general guide to achieve the design objective(s). It is vital that the designer realize the complexity of this technology, and that the many elements of design have symbiotic relationships that influence the synergistic effect affecting the quality of the effluent.

1. Wind directions and pond orientation.
2. Influent structure to pond.
3. Effluent structure from pond.
4. Hydraulic retention time in pond.
5. Depth of pond.
6. Wastewater characteristics to pond (influent).
7. Wastewater characteristics to filter (effluent from pond).
8. Wastewater characteristics of influent and effluent from other treatment units.
9. Wastewater characteristics of effluent from filter.
10. Anticipated algae from pond.
11. Organic loading on pond.
12. Operating conditions in pond-additional treatment in ponds. (aeration)
13. Anticipated effluent from filter.
14. Length x Width of filter. (Length is direction of flow)
15. L to W ratio of filter.

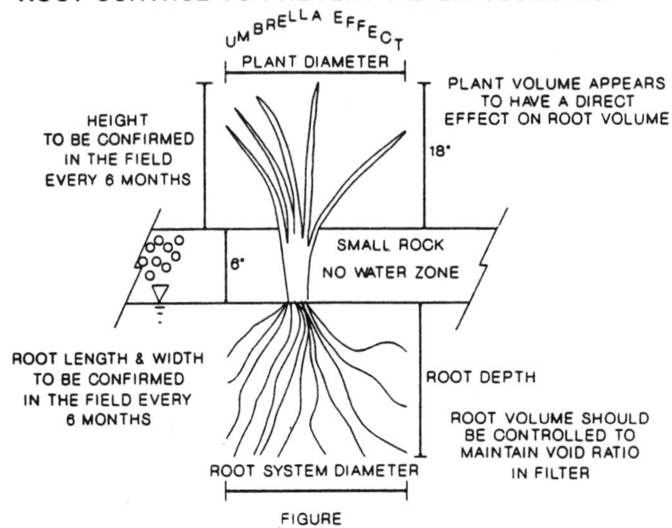

Figure 17. Root Control to Prevent Filter Clogging

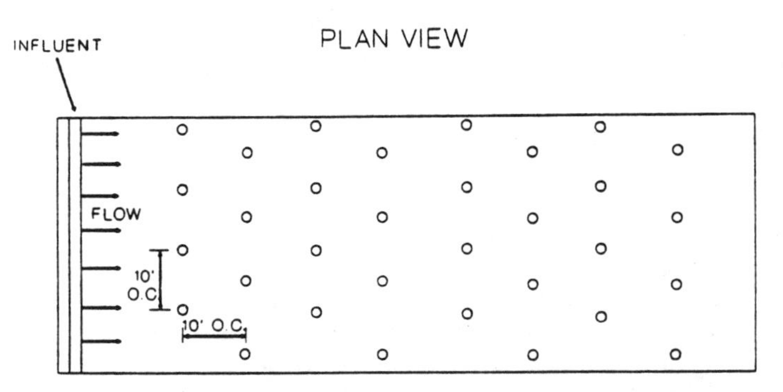

Figure 18. Planting Pattern for Filter

16. Influent structure to filter.
17. Effluent structure from filter.
18. Depth of filter.
19. Hydraulic retention time in filter.
20. Type of plants.
21. Number of plants.
22. Density of plants. (Super structure and root mass)
23. Planting pattern of plants.
24. Hydraulic gradient in filter.
25. Operating conditions in filter.
26. Rock gradation.
27. Rock sizes.
28. Depth of small rock. (no water zone)
29. Depth of large rock in filter.
30. Slope to bottom of filter.
31. Slope of filter surface.
32. Control of ponding in filter.
33. Temperature.
34. Allowance for evapotranspiration losses in filter.
35. Allowance for evaporation losses in pond.
36. Allowance for low flows.
37. Allowance for stormwater flows, infiltration/inflow analysis.
38. Provision for recirculation.
39. Provision to drain system. (Completely)
40. Removal procedures for undesirable plants.
41. Harvesting procedures for filter plants.
42. Removal procedures for excess filter plants.
43. Location of monitoring stations.
44. Design of monitoring station to ensure representative sample.
45. Protection against bank erosion.
46. Protection against leakage.
47. Allowance for insect and rodent control.
48. Acres required.
49. Design Flow. (Minimum, Average, Maximum)
50. Control of algae in ponds.
51. Control of algae in filter.
52. Control of root mass to maintain void ratio to ensure design detention time and subsurface flow in filter.
53. Control height of plants to achieve design objectives.
54. Establish criteria for selection of plants.
55. Conduct pan evaporation test at each site to determine evaporation and transpiration losses.
56. Determine plant transpiration rate.
57. Procedures to conform in-place void ratio.
58. Consideration of the use of calcium to enhance plant capability to uptake ammonia nitrogen.
59. Establish procedure to confirm actual detention time.

Design Objective(s). Before field investigation and review of existing records, objectives should be clearly understood. Objectives may be, but are not limited to: 1) achieve water quality standards, 2) reduce quantity of water, 3) reuse, and 4) conservation of energy and water.

Plant Criteria. Depending upon design objectives, different plants have different capabilities to satisfy design objectives to function at different sites in a specific wastewater. The filter void ratio and rock size will require a specific design adjustment. The void ratio of the filter rock will require adjustment to satisfy specific plant root substructures (volume) in order to maintain design flow at the design detention time.

Evaporation/Transpiration. Pan evaporation tests should be conducted at each site. From the pan evaporation test, compute the plant transpiration loss and the pond evaporation loss.

Recirculation Provision. Recirculation is vital during periods of low flow, high pond evaporation, and plant transpiration. It is also important when backflushing is necessary. Recirculation can increase detention time for additional treatment. Recirculation enables alternate modes of operation and provides operation flexibility.

Drainage Provision. Complete drainage is required for satisfactory backflushing. It is necessary for repairs to filter media and distribution system. Drainage is vital for measuring flow, water volume of filter, detention time in filter and provides data to measure the in-place void ratio.

Measure the In-place Void Ratio. Measure the in-place void ratio by filling the filter before planting. This will establish the in-place void ratio after construction is complete and before planting.

Measure Detention Time. After measuring the in-place void ratio, measure the detention time before planting. Measure detention time every 6 months until detention time remains constant.

In-place Void Ratio Adjustment. After confirming the void ratio of the rock media in place, plant the plants at a spacing that is to be maintained. Monitor height of plant and spacing and volume of root space. Predetermined volume must be assigned for root space for a specified spacing of the plants. For each operating spacing and assigned volume for roots, the filter will have an operating void ratio that must be monitored and measured every 6 months until the void ratio is a constant. Some plants may require removal to maintain required operating void ratio.

Measure Dissolved Oxygen. Measure the dissolved oxygen (DO) at the same frequency as other NPDES parameters, at the top, mid-depth, and bottom of filter. The DO should be measured at the influent, mid-point, and effluent.

Design Criteria from Developed Technology May Be Suitable for Technology Transfer under Certain Conditions. Design criteria should not be transferred from one site to another without first determining that site conditions are similar enough for such a transfer. Changing design criteria without performance data to substantiate the predicted result from the change is a risk that is not recommended. Without performance data, upon which to base a prediction, the anticipated effluent quality is in question. It follows, then, that the results of the technology transfer would be in doubt.

How to Determine Acceptability for Design Criteria Transfer.

- The size of the principal unit processes and operations must be such that physical, chemical, or biological processes will be accurately duplicated.
- All recycle streams have been considered.
- Process variables experienced will be the same as evidenced by accurate measurement.
- The time of testing has been adequate to ensure process equilibrium.
- Variations in influent characteristics substantially affecting performance have been accurately measured.
- Type and amount of all required process additives have been determined.
- The service life of high maintenance or replacement items has been accurately estimated from past performance.
- Full control of all major process variables has been substantiated by performance.
- Basic process safety, environmental, and health risks have been considered and found to be within local, state, and federal regulatory limits.
- All operational and management practices have been determined.

Risk versus Potential State-of-the-Art Advancement. Implicit in this objective is a willingness to accept a greater degree of risk in order to achieve a greater potential for a significant advancement in state-of-the-art as evidenced by lower cost, greater reliability, or other similar design objectives.

Construction Cost. The average construction cost for 22 facilities is $1.72/gal ranging from $0.62 to $3.62 per gallon. Cost includes site preparation and construction of pond and filter. The facilities varied in size from 0.26 mgd to 3.3 mgd. For locations and status see Figure 19.

Operation and Maintenance. For three facilities for 1989, the O&M costs were $.08, $.07, and $.10 per thousand gallons. Reported cost for 1990 at Benton is $0.16/1000 gallons.

Performance. Effluents for BOD, TSS, and NH_3-N ranged from 0-27, 5-30, and 1-10 mg/l, respectively, for six operating facilities in Louisiana.

Technology Assessment. We are still on the learning curve. The technology is still in the development stage. This is a success story. It is a viable technology especially for small communities, individual, and onsite systems. We do not have all the answers, but we do have sufficient design criteria based upon performance data to continue design, construction, and development.

Problems In Design. Some problems in design that remain include:

- How to control plant density to maintain detention time
- How to control growth of plant roots to maintain void ratio
- How to maintain plug flow and aerobic conditions
- Criteria for selection of plants

Sustainable Development. This is a sustainable development. Sustainable development is development that meets the needs of the present without compromising the ability of future generations to meet their own needs (7). A similar definition is "growth based on forms and processes of development that do not undermine the integrity of the environment on which they depend (8)."

LOCATION AND STATUS OF THE MRPF IN REGION 6 (01-31-91)
FIGURE 16

	AR	LA	NM	OK	TX	REGION
UNDER CONSIDERATION	0	10	0	1	3	14
PLANNING	5	8	2	1	0	16
DESIGN	8	2	0	0	4	14
UNDER CONSTRUCTION	1	5	0	5	0	11
OPERATING	0	25	2	0	2	29
TOTAL	14	50	4	7	9	84

Figure 19. Location and Status of MRPF in Region 6

A Challenge to Central and Regional Systems. There are times when Regional facilities are the most cost effective, but this technology is a real competitor and will challenge your imagination and ingenuity in solving our environmental problems. It will treat and reclaim air and wastewater to a quality for reuse at the point of waste generation. The environmental need must become a fundamental component rather than a constraint to economic development.

Innovation via Imagination. Have a desire and do not be afraid to develop alternative and innovative technologies beyond those found in textbooks, and do not be afraid to be imaginative and creative.

REFERENCES

1. Wolverton, B.C., and R.C. McDonald. 1975. "Water Hyacinths and Alligator Weeds for Removal of Lead and Mercury from Polluted Waters." NASA Technical Memorandum, TM-X-72723.

2. McDonald, R.C. 1981. "Vascular Plants For Decontaminating Radioactive Water and Soils." NASA Technical Memorandum, TM-X-72740.

3. Wolverton, B.C. 1975. "Water Hyacinths for Removal of Cadmium and Nickel From Polluted Waters"." NASA Technical Memorandum, TM-X-72721.

4. Wolverton, B.C., and R.C. McDonald. 1975. "Water Hyacinths and Alligator Weeds for Removal of Lead and Mercury from Polluted Waters." NASA Technical Memorandum; TM-X-72723.

5. Wolverton, B.C., and R.C. McDonald. 1977. "Wastewater Treatment Utilizing Water Hyacinths (Eichhornia Crassipes)." (Mary) Solms. pp. 205-208. "In Treatment and Disposal of Industrial Wastewaters and Residues." Proceedings of the national conference on treatment and disposal of industrial wastewaters and residues, Houston, Texas.

6. Wolverton, B.C., and D.D. Harrison. 1973. "Aquatic Plants for Removal of Mevinphos From the Aquatic Environment." J. Ms Acad. Sci., 19:84.

7. "World Commission on Environment and Development. 1987. Our Future. New York: Oxford University Press, 43.

8. MacNeil, P. "Strategies for Sustainable Economic Development," Scientific American, 2613:155-165 (September).

BIOLOGICAL NUTRIENT REMOVAL

Biological Nutrient Removal

Glen Daigger

CH_2M Hill
Denver, Colorado

<u>Overview</u>. This presentation covers the biological nutrient removal process options that are currently available and applied by the sanitary engineering profession. Systems for the removal of nitrogen *or* phosphorus and both nitrogen *and* phosphorus will be reviewed and compared. The operating characteristics of each system will be discussed and situations where each process might be used will be identified.

The presentation is structured to focus on the following concepts:

- An understanding of the mechanisms used in biological wastewater treatment systems to remove nitrogen and phosphorus
- The ability to apply the mechanistic understanding described above to analyze the capabilities of systems to remove nitrogen and phosphorus
- An understanding of the differences among the various biological nutrient removal process options that are currently available
- The ability to use the knowledge obtained from the understanding of the biological nutrient removal process options to select the process option most appropriate for a particular application

A BNR System is:

- **A Conventional Suspended Growth (Activated Sludge) Biological Treatment Process;**
- **Designed to Nitrify;**
- **With an Anoxic Zone Added for Nitrogen Removal;**
- **And an Anaerobic Zone Added for Phosphorus Removal.**

BNR System Reactors Are Segmented Into Anaerobic, Anoxic, and Aerobic Zones

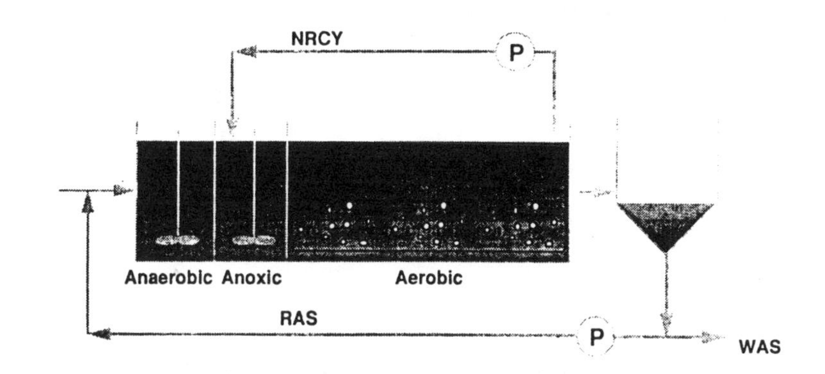

The Aerobic Zone Provides Nitrification and Final Effluent Polishing

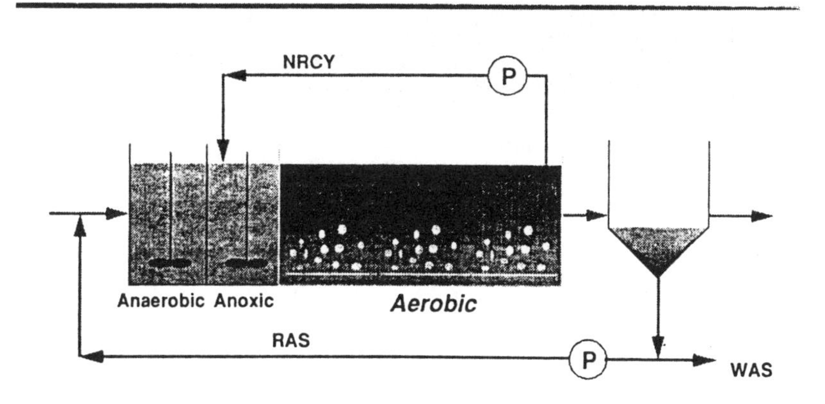

The Anoxic Zone and Nitrified Recycle (NRCY) Provide Denitrification Capability

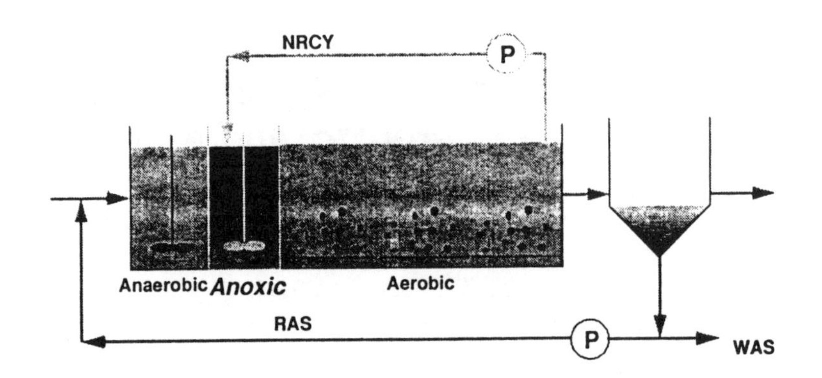

The Anaerobic Zone Provides Phosphorus Removal

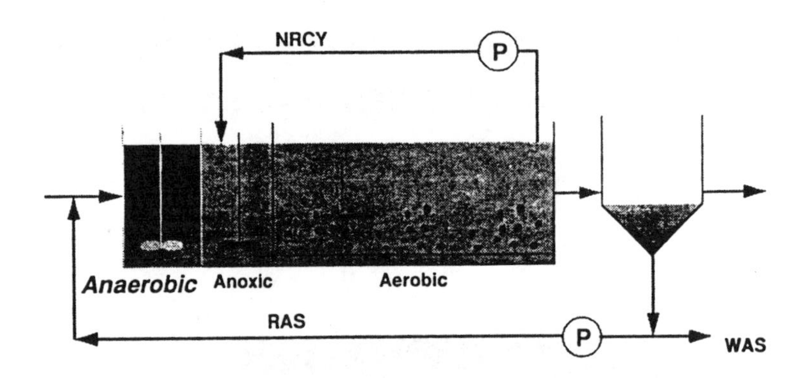

A BNR System will:

- Provide Nitrogen Removal
- Provide Phosphorus Removal
- Improve Sludge Setteability
- Reduce Process Energy Requirements
- Reduce Process Alkalinity Consumption

Nitrogen Removal Occurs Through Nitrification/Denitrification

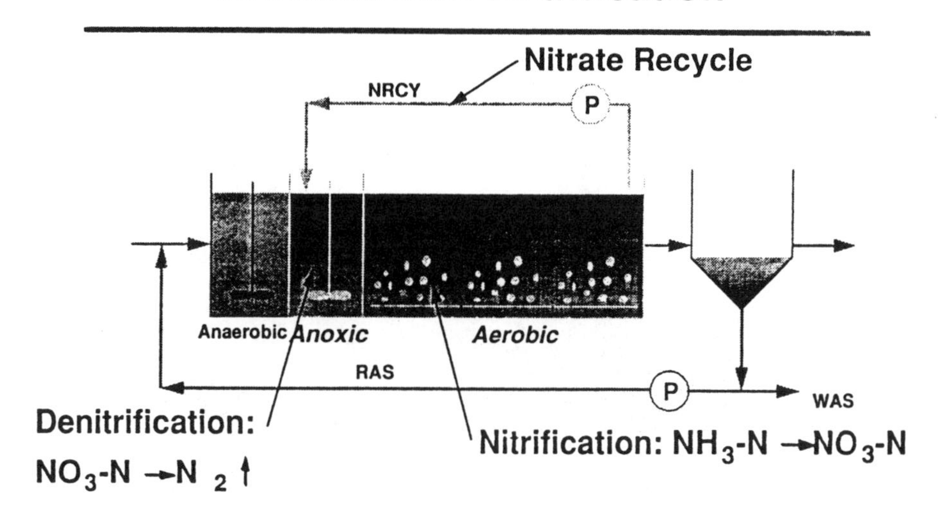

The Anaerobic Zone Selects for High Phosphorus Content Microorganisms

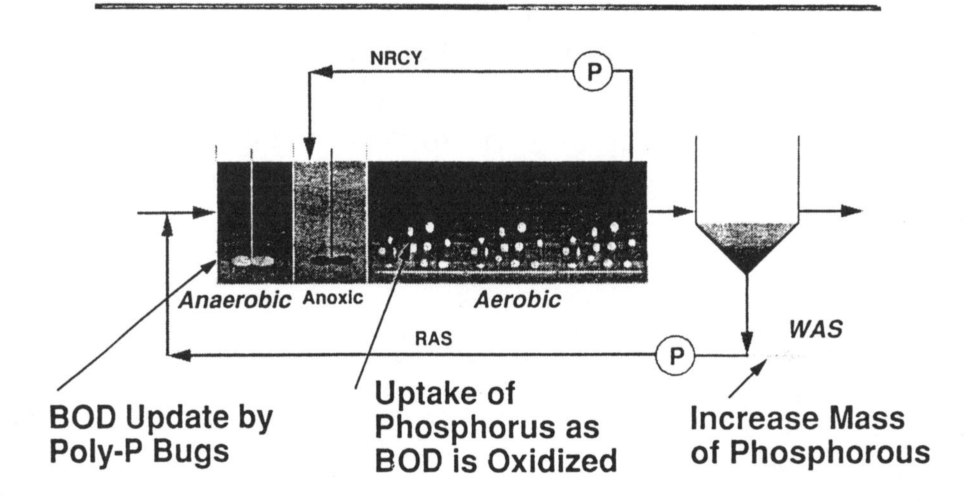

Sludge Settleability is Improved Because Filament Growth is Discouraged

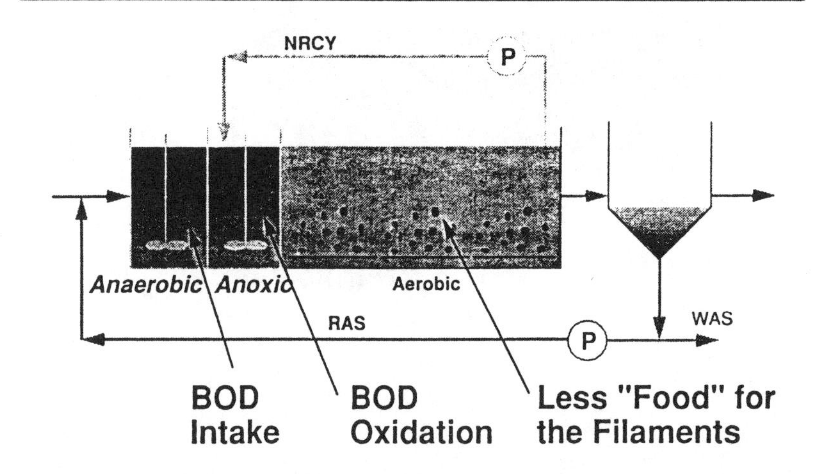

Complimentary Reactions Reduce Process Oxygen and Alkalinity Requirements

- Nitrification
  - 4.6 lb. O_2 *Consumed*/lb. NO_3 -N Generated
  - 7.2 lb. Alkalinity as $CaCO_3$ Consumed/lb. NO $_3$-N Generated
- Denitrification
  - 2.86 lb. Oxygen Demand *Satisfied*/lb. NO_3 -N Removed
  - 3.6 lb. Alkalinity as $CaCO_3$ Produced/lb. NO $_3$-N Removed

VIP Project Objectives Emphasized Cost-Effective Level of Nutrient Removal

- Existing Plant Upgrade
- Total Process HRT of 6.5 Hours
- Remove Two-Thirds of Influent Phosphorus
- Remove 70% of Total Nitrogen Seasonally
- Obtain Associated Process Benefits
  - Oxygen
  - Alkalinity

VIP Process Uses Anoxic Recycle (ARCY), Staged Configuration and High-Rate Operation to Optimize Phosphorus Removal

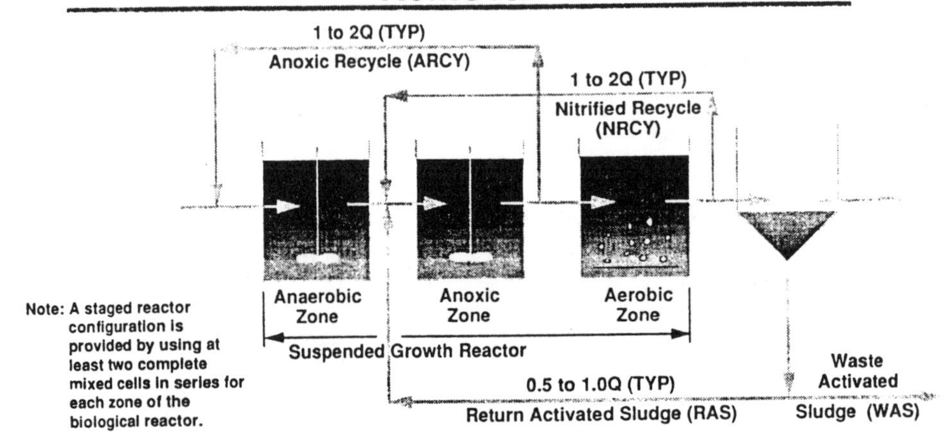

The 40-mgd VIP Project Began in 1985 and is Nearing Completion

- **Facilities Plan Ammendment**
- **15 Months of Pilot Testing Verified That Project Objectives Achievable**
- **18 Months of Successful Full-Scale Operation at York River**
- **Several Other Pilot Plants Completed**
- **40-mgd VIP Plant Start-Up in May, 1991**

Operation of Anoxic Selector Activated Sludge Systems for Nitrogen Removal at Rock Creek and Tri–City Wastewater Treatment Plants

Gordon A. Nicholson

CH_2M Hill
Portland, Oregon

Overview. The Rock Creek and Tri-City treatment plants were the first facilities in the Pacific Northwest to have anoxic zones incorporated into the designs of the aeration basins. Anoxic zones aid in controlling filamentous growth and the removal of nitrogen. Mixed liquor is recycled to the head of the aeration basin where under anoxic conditions nitrate is used as the electron acceptor for the uptake of soluble BOD. Metabolization of the BOD occurs in subsequent aerobic zones of the aeration basin. Besides removing nitrogen and controlling filamentous organisms, the anoxic selectors aid in the control of pH and reduce aeration demands. The Tri-City plant has been operating for 5 years, consistently producing a high-quality effluent and maintaining SVIs less than 100. The Rock Creek plant has produced similar results with 2 years of operational history. Design criteria, operational performance, and differing control strategies are included in this presentation.

The Objectives Of This Presentation Are:

1. **To Update Anoxic Selector Performance at Tri-City**
2. **To Describe Anoxic Selector Performance at Rock Creek**
3. **To Compare Performance of the Two Systems**

Anoxic Selector Systems Possess Several Documented Advantages

- ☐ Control of Filamentous Organisms
- ☐ Nitrogen Removal
- ☐ Reduced Alkalinity Consumption
- ☐ They are also Non Proprietary

Anoxic Selectors Are Provided At Both Tri-City And Rock Creek

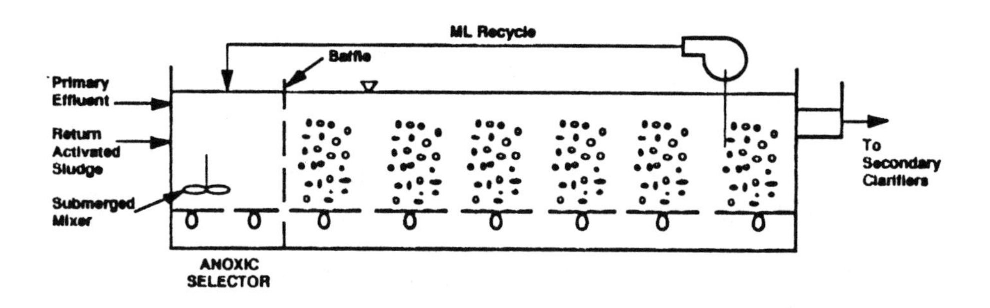

At Current Loadings One Half The Facilities Are Operated At Tri-City

	Aeration Basin HRT (hr.)	Overflow Rate (gpd/ft²)
Ave	5.3	508
Max	8.3	891
Min	3.5	326

Effectiveness Of Anoxic Selector In Controlling SVI Demonstrated At Tri-City

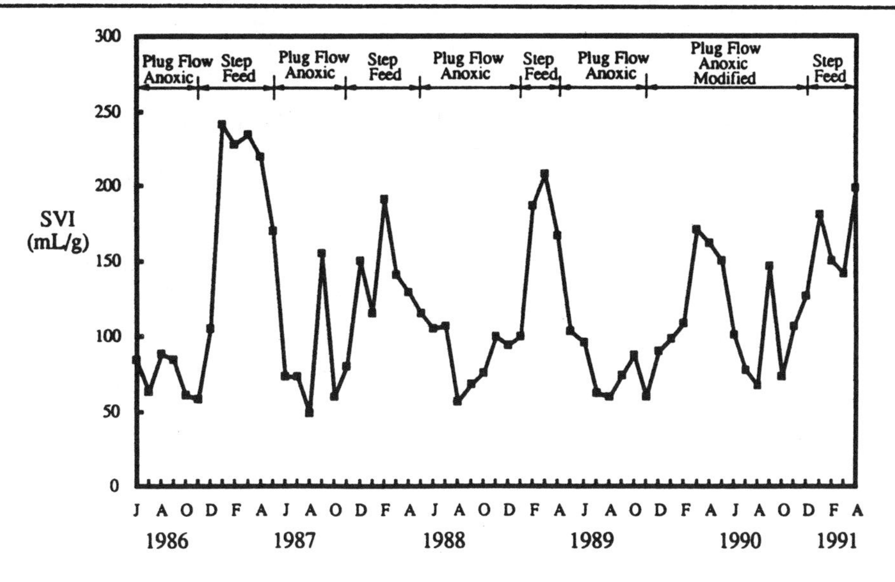

Effluent Quality Is Reliable Below 20/20 At Tri-City

	30 Day Average (mg/L)	
	BOD	TSS
Ave	7	8
Max	19	20
Min	2	2

Nitrogen Removal Is Maintained Year Round At Tri-City

	30 Day Average (mg-N/L)		
	NH3	NOx	TN
Ave	2.0	7.0	9.0
Max	8.1	11.1	13.1
Min	0.3	4.1	4.8

Process Loadings Are Similar At Rock Creek

	Aeration Basin HRT(hr)	Overflow Rate (gpd/ft^2)
Ave	5.7	513
Max	8.5	580
Min	4.7	349

Primary Effluent Characteristics Vary Seasonally At Rock Creek

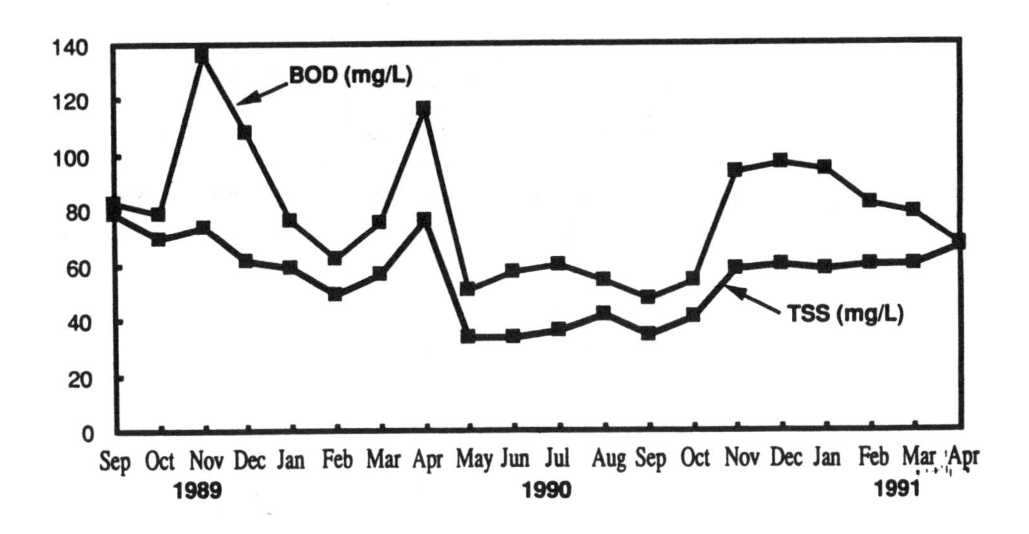

Nitrogen Removal Varied Seasonally At Rock Creek

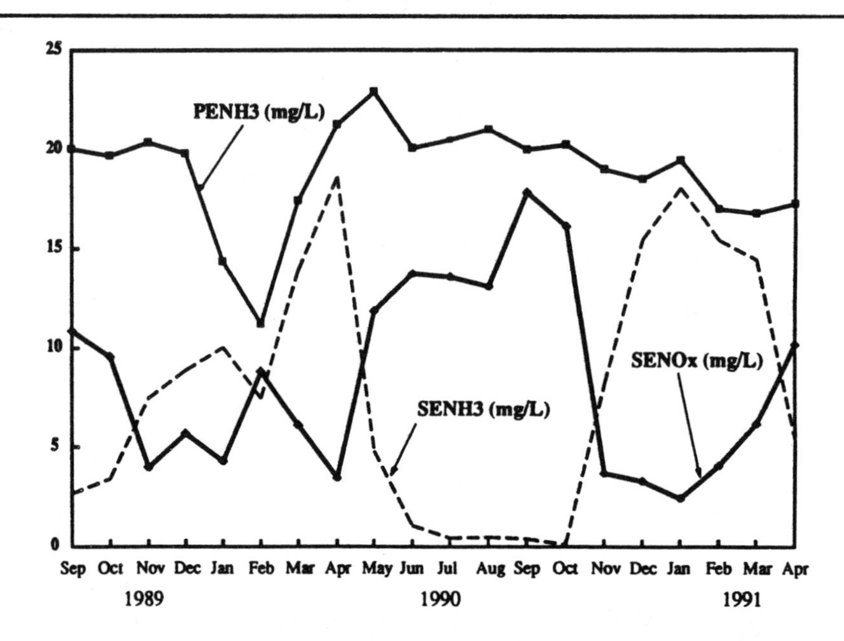

SVI Also Varied Seasonally At Rock Creek

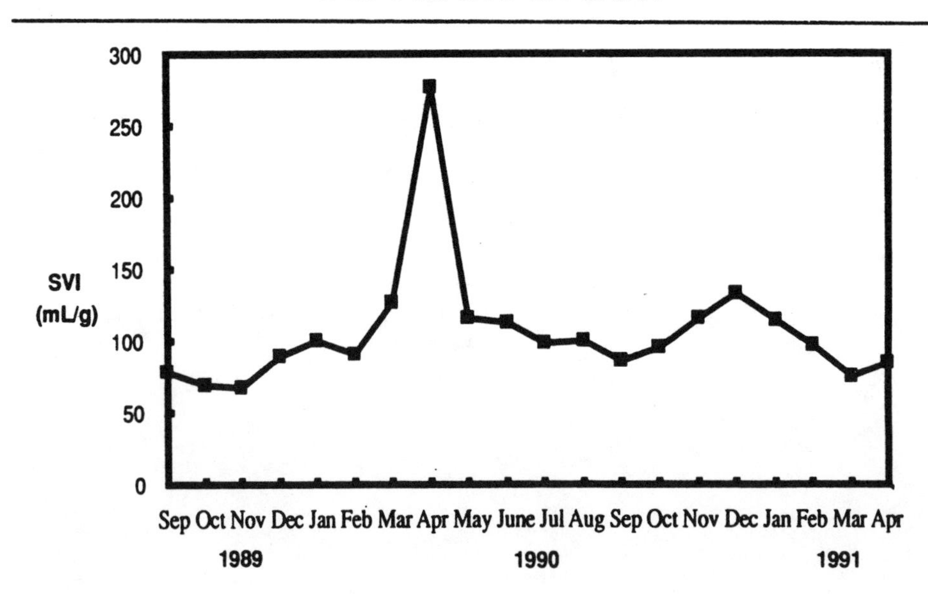

Secondary Effluent Quality Has Been Excellent At Rock Creek

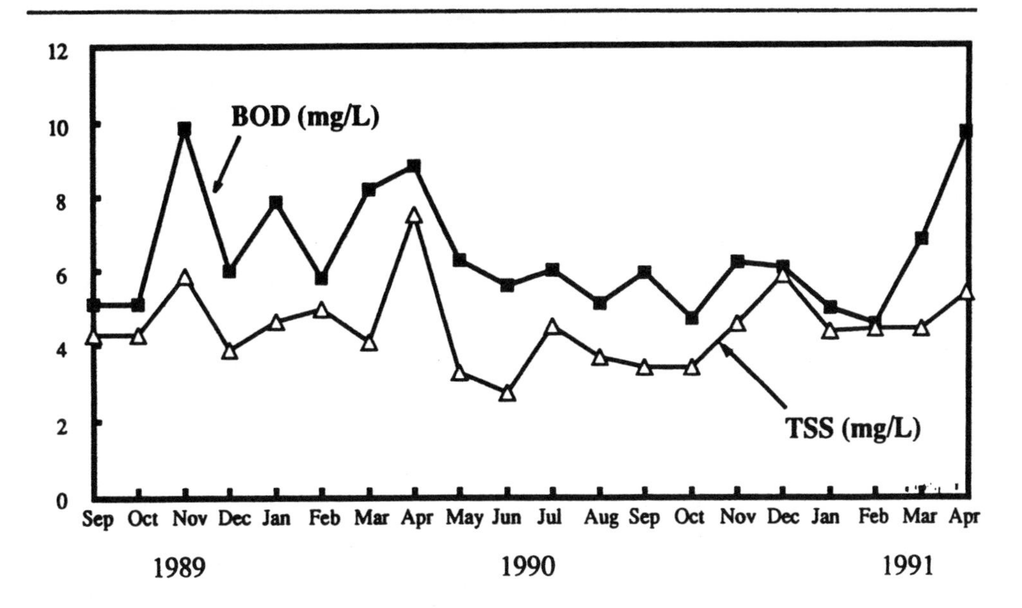

Experience At Tri-City And Rock Creek Demonstrates That:

1. **Anoxic Selectors Effectively Control Filamentous Bulking**
2. **Nitrogen Removal and Alkalinity Recovery are Functions of BOD Loading**
3. **Design Details Facilitate Scum Handling**
4. **Effluent Quality is Good**
5. **Effluent BOD/TSS Varies between Facilities**

Summary of Patented and Public Biological Phosphorus Removal Systems

William C. Boyle

University of Wisconsin
Madison, Wisconsin

The removal of phosphorus from municipal wastewaters to control receiving water eutrophication has been receiving high priority in many states and may become a significant constraint in the NPDES discharge permit of many municipalities. Technologies exist for removing phosphorus by physical, chemical, and/or biological means. Biological phosphorus removal (BPR) has rapidly emerged as a desirable alternative process because of its relative ease of implementation at existing plants using conventional activated sludge treatment.

There are a number of BPR system flowsheets in use today that are claimed to be patented, although in certain cases the validity of the claim is in question. EPA needs to establish which BPR processes lie in the public domain, which technical and/or other approaches to BPR are currently valid processes, and the judicial rulings/findings associated with selected BPR patent claims. The objective of this work was to summarize the status of patented and public BPR systems in the United States so that the technology will be better understood by the technical and regulatory communities.

This project consisted of two basic tasks:

1. Establish which patents issued since 1960 in the United States appear to cover BPR processes and identify claims and holder of each.

2. Contact federal and state agencies, design engineers, and others knowledgeable in BPR processes and identify municipal facilities where no license fees have been paid, no allegations of patent infringement have been made, and/or no lawsuits have been filed. These facilities are to be described, current or future NPDES permit requirements identified, and plant operation/performance data summarized for the past 12 months.

<u>**Patent Search — 1960-1990**</u>. A patent search on BPR processes was conducted by Christie, Parker, and Hale. They identified 50 patents since 1960 that involve biological treatment processes that in some way resulted in the removal of phosphorus. An additional 100 patents

were also identified that had aerobic and anaerobic treatment steps that inherently may result in some phosphorous removal, even though not specifically mentioned. A review of the 50 primary patents revealed several important patented flowsheets currently used by municipalities in the United States and a number of concepts that do not appear to be directly applicable to or practical for municipal processes. Table 5 lists some of the currently licensed patents, assignee, and number of facilities in operation or design/construction.

A search for litigation involving the 50 patents disclosed only one lawsuit by Air Products & Chemicals vs. Orange Water and Sewer Authority in the Middle District of North Carolina, filed January 6, 1988. It appears that this suit has now been withdrawn, but further details are still unavailable. One other point of interest on these patented processes is the period of patent coverage: 17 years. Any reissue or "improvement" on the process reverts back to the date of original patent issue, although specific improvements are extended 17 years from their disclosure.

BPR Processes in Public Domain. A telephone survey of all states in the United States attempted to identify municipal BPR processes currently operating or under design/construction that are in the public domain. Needless to say, this is an enormous task and the likelihood of omissions is high, especially for small facilities. Table 6 presents the tentative results of this survey whereby generic system names are used to identify processes. Again, the numbers cited for each system are tentative, at best.

A tentative review of the performance of public systems now in the ground suggests substantial variability in process results. Most SBR and oxidation ditch systems employ chemical additions to "polish" final effluent phosphorus levels. Most have been in service less than 2 years and manufacturers/engineers are continuing to experiment and refine process operation. The VIP and UCT flowsheets, which are very similar, have received significant attention in the United States, and the process reliability appears to be very good based on pilot and demonstratlon experience. Primary sludge fermentation (PSF), which may be a side-stream or main-stream process, has been successfully applied at Kelowna, BC, and Orange Water and Sewer Authority (Chapel Hill, NC). It is currently being applied at two Bardenpho installations to upgrade performance. Licensing issues related to this process are yet to be resolved. Finally, operationally modified activated sludge systems are probably grossly underestimated in Table 6

Table 5. Summary of Selected Currently Licensed BPR Systems

Assignee	Trade Name	Facilities in Operation	Facilities under Design/Construction
Air Products	AO, A^2O	18	19
EIMCO	Bardenpho-5 Stage	22	17
Biospherics, Inc	Phostrip	7	?
Orange Water/Sewer	OWASA	1	?
Hampton Rds. San. Dist.	VIP	3*	4
Transenviro, Inc.	CASS	10	9

*Pilot and demonstration system

Table 6. Summary of Selected Currently Used BPR Systems in Public Domain

Generic System	Facilities in Operation	Facilities under Design/Construction
SBR	8	1
Oxidation	11	4
Operationally Modified A.S.	4	?
VIP	3*	4
VCT/Modifications	?	3
Primary Sludge Fermentation	2	1
Aquatic Systems	1	?

*Pilot and demonstration systems

owing to litigation concerns. Two factors may result in significant increases in the successful application of this process: phosphate detergent bans and the reissue of the AO patent (1987), which now specified "absence of supplied oxygen-containing gas" in the anaerobic zone.

<u>Future Outlook</u>. Results to date indicate that most mainstream BPR systems in operation can meet a 1.0 mg/L P requirement most of the time. Factors that dictate good performance include high soluble BOD to soluble P ratio (volatile fatty acids generated in mainstream or sidestream reaction); low DO and nitrate concentrations in anaerobic zone; high sludge yields; and effective P control in sludge processing. Sidestream processes such as Phostrip are generally able to achieve lower effluent P values because of their operational flexibility and insensitivity to influent wastewater BOD.

Selection of future BPR processes for greenfield or retrofit applications will be dependent on NPDES discharge permits and influent wastewater characteristics that may be greatly influenced by detergent phosphate bans. The currently licensed BPR processes will likely continue to be used in many instances because of the years of experience with these flowsheets. Several patents will lapse within a few years, a fact that is significantly influencing some engineers in process selection.

PSF processes provide significant flexibility to plants with weak wastewaters, those with fixed film processes and those having difficulty with high anaerobic zone nitrates. The concept provides the engineer with an opportunity to provide greater operational stability and reliability to mainstream BPR systems. One lingering question however relates to how the OWASA patent may affect new PSF designs. Early interpretations of the OWASA patent suggest that it applies to fixed film or chemical pretreatment processes where BOD/TP ratios are significantly reduced prior to the BNR system. A broader interpretation of this patent may occur as future designs are developed.

As discussed earlier, operationally modified activated sludge designs and retrofits may be pervasive in the future as BOD/TP ratios increase. Combined with VIP/UCT flowsheets, it is likely that engineers will select maximum flexibility with internal mixed liquor recirculation, multiple sludge recycle feed points, tank baffling, and even PSF sidestream applications.

Use of oxidation ditches and SBR configurations will likely continue for small facilities. More experience with system operation and process modifications is in order and currently underway. In many system designs, especially where NPDES permits may become more stringent in the near future, standby chemical feed is an intelligent choice to ensure that the municipality meets its permit under adverse situations. Effluent requirements below 0.5 mg/L may dictate filtration polishing for most currently used BPR systems.

SLUDGE

Case Study Evaluation of Alkaline Stabilization Processes

Lori A. Stone

Engineering-Science, Inc.
Fairfax, Virginia

The management of sewage sludge continues to be a major problem for municipalities. Municipal sewage sludge management programs are being influenced by a number of factors, including: increasing sludge volumes resulting from better wastewater treatment; the U.S. Environmental Protection Agency's (EPA's) "Beneficial Use Policy," which actively promotes the beneficial use of sludge while maintaining or improving environmental quality and protecting public health; current and proposed federal and state regulations; conversion of the Construction Grants program to a state revolving fund program; public acceptability; and a desire by municipalities to select a cost-effective sludge management option. As environmental issues continue to influence sludge management programs, municipalities are becoming aware of alternative sludge use practices.

New forms of chemical stabilization other than lime treatment have been developed by vendors and are being used by municipalities. These technologies add alkaline materials such as cement kiln dust, lime kiln dust, Portland cement, or fly ash for the stabilization of sludge. Most of these technologies are modifications of traditional lime stabilization. The most common modifications include the addition of other chemicals, a higher chemical dose (depending on the chemical type), and supplemental drying. These processes alter the characteristics of the sludge and, depending on the process, increase the stability and physical strength, decrease the odor potential, and/or reduce pathogens.

As more municipalities become aware of these newer, alkaline stabilization technologies, there was a need to compile information from operating, full-scale projects. An evaluation was conducted to determine:

- How the major alkaline stabilization process have been and are being used
- The problems that have been encountered

- The solutions that have been implemented to improve the processes

Site visits to four alkaline stabilization facilities were conducted in addition to follow-up meetings with vendor representatives, plant managers, and regulatory officials. The four facilities visited were: New Haven, Connecticut; Salem, Massachusetts; Toledo, Ohio; and Greenville, South Carolina. The New Haven and Salem operations use the Chemfix process; while the Toledo and Greenville facilities use the N-Viro process. Although the Chemfix and N-Viro processes were the focus of this study, similar technologies have emerged which also use alkaline additives to stabilize municipal sewage sludge, such as the BioFix IV process of BioGro Systems and the EnVessel Pasteurization process of RDP, Inc. This document is not intended to single out one design or technology as superior to another.

Case study evaluations were prepared to summarize the information gathered and the observations made during the site visit. It should be emphasized that each case study represents a "snapshot" of the process and operations at the time of the visit. The factual data presented, such as the loading rates, operating practices, and odor production and control, are accurate only for the time of the site visit and may have subsequently changed because the systems are still rapidly developing.

The information collected from these visits and conferences is intended to assist municipalities that are considering these processes in their own site-specific assessments. Specifically, this document focuses on important issues that should be considered when evaluating alkaline stabilization as a sludge management option. These issues include project planning and implementation considerations, associated costs, procurement options, process operations, monitoring and regulatory requirements, and product quality and use.

The study found that the N-Viro and Chemfix processes can be implemented in a very short period of time. In many cases, these alkaline stabilization processes have been used to solve an immediate problem (e.g., the shut down of another system or an odor problem) and then for various reasons have been replaced by the repaired or other system. In addition, several full-scale systems using alkaline processes have and are being implemented as long-term solutions. Available information about these systems is summarized in Table 7.

Table 7. Comparison of Alkaline Stabilization Processes

Comparison Category	ChemFix	N-Viro
Number of Facilities[1]	9	19
Percentage PSRP[2]	100%	26%
Percentage PFRP[2]	0%	74%
Average Dry Tons of Sludge Processed per Day	45	12
Chemicals Typically Used in Sludge Stabilization	Type I Portland Cement Sodium Silicate	Cement Kiln Dust (CKD) Lime Kiln Dust Quicklime, Pulverized and/or Hydrated Lime
Predominant End-Use for PSRP Product[2]	Landfill Cover Applications	Land Application/Soil Conditioner (45%) Other[3] (55%)
Predominant End-Use for PFRP Product[2]	No PFRP Facilities Currently Exist	AgLime/Soil Conditioner (88%) Landfill Cover (9%) Other[3] (3%)
Typical Procurement[4]	Privatized (100%)	Privatized (17%); Owned by Municipality/Operated by Third Party (33%); Owned and Operated by Municipality (50%)

1 Includes pilot and demonstration facilities.
2 Based on DTPD weighted average of all operational facilities.
3 Includes berming, diking, , and privately owned yard development.
4 Privatized procurement includes ownership and operation of the facility by a private firm. If the process is owned and operated by the municipality or third party, a technical service and licensing fee may be charged. The percentages given above for the N-Viro procurement methods are based on available information from 12 operating facilities.

The planning and implementation of a sludge management option depends on several factors. Depending on the proposed end-use for the final product, alkaline stabilization can be an effective management option for sludges with varying degrees of sludge quality. The addition of alkaline, pozzolonic material serves to dilute concentrations of heavy metals and organic chemicals on a dry weight basis and tends to immobilize these constituents. Although the Chemfix and N-Viro processes could be modified to treat hazardous sludges, both vendors require that the sludge be non-hazardous so the final product can be used in landfill or agricultural applications.

From the study, it is apparent that the feed sludge and alkaline material should be monitored frequently so that adjustments could be made to the process to maintain a consistent product. Because the quality of the alkaline additives may have a direct effect on the quality of the final product, it is extremely important that adequate monitoring be performed to ensure consistent quality supply. More importantly, pilot or bench-scale testing should be performed to determine how variations in the alkaline additives will affect the final product quality and how the process and chemical dosages should be adjusted to account for these variations. Parameters to be monitored include total solids, pH, and temperature of both the sludge and final product. Quality control data were required for regulatory approval at each of the facilities visited. The method and frequency at which these parameters were monitored depended on the regulatory requirements. In some cases, odor characterization and emission monitoring was required.

The product end-use has particular quality requirements and standards associated with it. Depending on the end-use, it may be necessary to satisfy either Process to Significantly Reduce Pathogens (PSRP) or Process to Further Reduce Pathogens (PFRP) requirements. Potential markets for an alkaline stabilized product include agricultural, slope stabilization, structural fill, or municipal landfill cover operations. As shown in Table 7, the majority of the N-Viro facilities are PFRP; all of the Chemfix operations have been PSRP. Accordingly, the predominant end-use for the N-Viro product is as an agricultural liming agent or soil conditioner, which typically requires PFRP treatment; whereas, the Chemfix product is used mainly as cover material for landfills, requiring only PSRP treatment. Research studies have been performed investigating the use of the Chemfix material as a soil conditioner and as fill material. However, to date, the Chemfix product has been used only in landfill operations for grading purposes and as daily, intermediate, and final cover.

Although these technologies use similar materials handling equipment as traditional lime stabilization, the precise chemical formulas of the stabilization additives or processing steps are generally proprietary. Consequently, many of these technologies are only available to municipalities only as procurements from private firms. The type of procurement and associated economics are important issues to consider when selecting a sludge management option. The costs of an alkaline stabilization process should be evaluated with respect to the total costs of the option over its useful life using present worth, equivalent annual cost, or similar method. In addition, the costs of a privatized option, if that is the preferred procurement method, should be compared to alternative management options that are to be owned and operated by the municipality. The privatized facility is designed, operated, constructed, furnished with all the necessary equipment, and eventually operated as a commercial enterprise by a private firm. As shown in Table 7, the Chemfix facilities are privatized procurements. Chemfix will allow municipalities to operate the process using their own personnel and facilities; however, a licensing fee for the use of the process may be charged. Although the majority of the N-Viro facilities are owned and operated by the municipality, a royalty fee is charged for the use of the patented technology. Several of the N-Viro facilities are owned by the municipality and operated by a third-party through a licensing agreement. A technical service fee may also be applicable for the initial licensing of a facility. This fee includes items such as process design, pilot testing, review of drawings and specifications, facility startup services, and initial PFRP compliance testing.

Flexibility (adaptability) of the alkaline stabilization process and the use of existing facilities should be considered when evaluating potential sludge management options. Although it may be not always be possible to use or retrofit existing equipment, cost savings can be achieved by the municipality if existing equipment is used as part of the alkaline stabilization process train. Existing equipment was used to minimize capital expenditures at several of the facilities visited. Depending on the site constraints, the type of process utilized, and the amount of sludge to be processed, site preparation for both the Chemfix and N-Viro processes is generally minimal. The process equipment can usually be arranged to accommodate various site constraints. Because these processes are relatively simple to operate and do not require extensive, complex processing equipment, they can usually be implemented in a short time frame and require relatively minimal space. Additionally, both Chemfix and N-Viro have mobile, skid-mounted equipment that can be used for back-up or emergency situations and demonstration programs to encourage interest in the final product.

Since the processes are not manpower or equipment intensive, alkaline stabilization can be a cost-effective option depending on the product market and end-use program. In addition, an advantage of these processes is the ability to start-up operations in a relatively short time period. Because of their flexibility and ease in start-up, alkaline stabilization can be an effective sludge management option, especially as an interim alternative.

Controlling Sludge Composting Odors

William G. Horst
City of Lancaster
Lancaster, Pennsylvania

and

Bert deVries
Connert, Inc.
Lancaster, Pennsylvania

In the early 1980s Lancaster began planning and designing new wastewater treatment facilities. The City had a long history of odor complaints from the neighbors. Since the Zimpro process for sludge handling was the cause of most of the bad odors, it was decided to look for an alternate sludge handling process. To fully understand the odor problems, it should be pointed out that the topography and location of the plant put neighbors in proximity with these undesirable smells. This presented us with a major difficulty in arriving at a solution. The plant is located in the lowest area of the city and the neighbors above are looking down onto the sewer plant and the malodorous fumes are rising directly toward them with little chance of dispersion.

The Authority therefore searched for a method to contain odors. What better way to achieve this objective but to use "in-vessel" composting. This appeared to be the answer.

However, it was immediately apparent that "fugitive" emissions were a major problem. Open conveyors on top of the Taulman Weiss composting vessels allowed unobstructed odor release during many hours of each day. Also, the open top feed apertures of the vessels allowed odors to escape all day long.

The City decided to tackle the fugitive emissions first. All conveyors were enclosed and the air within the conveyors ducted to a Quad mist scrubber. This effectively controlled these sources.

The second major construction job was to greatly increase the suction out of all four reaction vessels. Three to four times more air was taken out of the vessels than aeration air was blown into the vessels. That created a suction inside the vessels. In fact, vacuum breakers were installed to prevent a collapse of the tops of the vessels. This excess exhaust air prevents air from escaping and becoming "fugitive" odors.

As part of the process instructions by Taulman Weiss, the City also turned off all aeration blowers while loading or unloading vessels and while transferring material from one vessel to another. Up to 8 hours each day, therefore, the biological process received no aeration. Upon resumption of the aeration blowers, a large "spike" of odorous air, often at very high temperature, would be blown into the odor control system. There was really no way to control this variable "odor load."

Because of the higher airflow, and to somewhat "dampen" the odor fluctuations, the Authority placed two scrubbers in sequence to scrub the odors.

The Pennsylvania Department of Environmental Resources last year accepted recommendations to allow the City to work out a system to chemically treat the odors rather than use a very costly incinerator. To that end Bert deVries, formerly with Quad, but now an independent consultant, was hired to coordinate and expedite the work in progress.

The first scrubber is a "water wash" scrubber, with the multiple purpose to:

1. Cool the hot odorous gases.
2. Neutralize ammonia by acidifying the water.
3. Dampen the "shocks" which occur after the aeration blowers are put back on.

The second scrubber is a Quad mist scrubber which removes carbon sulfides and the many other persistent odors.

Both scrubbers are pH controlled since both ammonia reduction and the oxidation of the odors in the second scrubber take place under carefully controlled acidic conditions.

Readily available "process water" is used in the cooling-stripping scrubber and only a 1½ gallon per minute mixture of chemicals and potable water is used in the second scrubber.

Contrary to the Taulman-Weiss recommendation, we presently continue to aerate 24 hours a day. The excess air taken off at the top of the composting vessels creates a sufficient vacuum, so that it is now possible to continue operation of at least 1,150 cfm by the aeration blowers to supply air into the bottom of the vessels. The benefits of this procedure become immediately obvious. The temperature of vapors into the "cooling" scrubber no longer rises 20 to 30°F. The "spike" of sudden odor levels to be treated also disappears. The more gradual increases and decreases of temperature and odor levels can now be tracked automatically to a better degree.

All the above improvements to solve the technical problems will still require attention and care by the staff. It has become obvious once again, that good technology still depends on good operators and that constant vigilance will be required to keep our neighbors "happy" neighbors.

The Authority also is grateful for the patience and encouragement of the PADER who stood by us throughout these difficult years. We understand the pressure they were subjected to by neighbors who wanted a "quick-fix."

The City feels confident that after all the seemingly insurmountable problems, a practical and innovative system was developed on-site. The City believes that others may benefit from our experiences and persistence so that composting as an alternative sludge disposal technique may shed some of its unfavorable image acquired over the last few years. This is new technology, and the City feels it contributed by the lessons learned.

Total Recycling

Dale Cap

Southwest Suburban Sewer District
Seattle, Washington

Background. Southwest Suburban Sewer District's first wastewater treatment plant became operational in 1956 providing primary treatment and chlorine disinfection. The Salmon Creek plant was followed by construction of the Miller Creek plant in 1965 which provided the same degree of treatment. Sludge was anaerobically digested, dewatered by a vacuum belt filter and then later by a centrifuge, and made available to the public. All sludge produced at that time (up to 750 yd^3 in 1986) was stored originally at the Salmon Creek plant and then later on at the Miller Creek plant and utilized by the public. The dewatered sludge had a solids content of 30 percent, was of a fairly loose texture that made it easy to handle, and the odor was not too strong.

Construction took place in 1986 through 1989 to upgrade both treatment plants to secondary treatment. Rotating biological contactors, a fixed film process, coupled with solids contact was selected as the biological treatment process, and digester capacity was expanded to handle the additional sludge. The biological solids changed the nature of the sludge throughout the process. It doesn't compact as well nor dewater as easily after digestion, and it has a strong, predominantly ammonia odor. This material, though providing the same or even greater benefits for soil improvement, wasn't desirable for the average homeowner to work with. Furthermore, the volume of dewatered sludge increased four- or five-fold.

Sludge Quantity. Each treatment facility treats a wastewater flow of about 3.5 mgd and pumps 11,000-13,000 gallons of 4 percent total solids raw sludge to the anaerobic digestion process daily. Volatile solids reduction averages 50 percent to 65 percent. Each plant produces about 28 yd^3 per week of 18 percent total solids dewatered digested sludge for a total of about 3,000 yd^3 per year. The majority of the sludge is composted on-site for distribution and marketing, and a small amount is applied to land.

Composting. The District considered building a compost facility to produce a marketable product that could increase the demand to equal the supply. Composting meets the EPA criteria for a Process to Further Reduce Pathogens (PFRP) because of the high temperatures reached in the process (above 55°C). After much investigation, including visits to installations across the country, the idea of an in-vessel composter appeared to be too costly and risky as far as proven reliability. The District decided to build a large odor-controlled enclosure that provided flexibility for static pile composting. The composting project is working very well. The composted product we are now producing is a rich-looking humus that is loose and very workable with virtually no odor. With the increasing interest being shown, we are selling everything we can produce. For our static pile composting operation we use a front loader with a 2 yd^3 bucket for all the materials handling. The sludge is mixed by volume, one part dewatered digested sludge to one part sawdust (or other amendment, such as yard waste) to one part recycled compost. The mixture is organized into piles or rows in the compost building, and numbered temperature probes are inserted into the pile to track the temperatures. Once the temperature has remained above 55°C for at least three days, the pile is remixed and aerated. We do this at least three times with the temperatures getting above 55°C to assure a safe reduction of possibly harmful bacteria. Once a batch has substantially composted so that there is no odor, it is moved out of the enclosed odor-scrubbed compost building into an open-sided curing building where it is available for sale to the public. People bring their own containers or pick-up trucks to purchase the compost at $10/$yd^3$ with a $2 minimum. We deliver 4 yd^3 minimum loads within a reasonable distance at no extra cost. We advertised our compost in the District's newsletter, which is sent to District customers periodically. In a relatively short period of time all the compost was sold and there was a waiting list for the next available batch.

We are fortunate that our treatment plants serve a "bedroom community" with little or no industrial waste so that we have a high quality sludge that can be beneficially and safely recycled into the environment. We started a program in cooperation with the City of Normandy Park for recycling yard waste into our compost. At present we are accepting only grass clippings, leaves, and other materials that don't require grinding. This has had a number of benefits. It has given us more free amendment, it has increased the public visibility and awareness of our program in a positive light, and it has put these valuable resources, sludge and yard waste, to work for us.

STORMWATER

Washington State's Approach to Combined Sewer Overflow Control

Ed O'Brien

Washington State Department of Ecology
Olympia, Washington

Introduction. Washington State has embarked upon a combined sewer overflow (CSO) control program whose goals are:

- To reduce the incidence of untreated overflows to an average of once per year per overflow site
- To comply with sediment quality standards and water quality standards during and after overflow events
- To not preclude unrestricted use of the receiving waters for their characteristic uses

The program is both technology- and water quality-based. The pace of its implementation is decided on a case-by-case basis by applying economic criteria to each municipality's situation. Figure 20 presents a schematic of Washington's CSO treatment requirements.

Overview of the Statute and Regulation. In 1987, Washington amended its Water Pollution Control Act to include provisions specific to the reduction of combined sewer overflows. Those provisions require the Department of Ecology, the state's pollution control regulatory agency, and local governments "to develop reasonable plans and compliance schedules for the greatest reasonable reduction of combined sewer overflows . . . at the earliest possible date" (RCW 90.48.480). The initial reduction plans and schedules were due by January 1, 1988.

To implement this legislation, the Department of Ecology adopted a new regulation, "Submission of Plans and Reports for Construction and Operation of Combined Sewer Overflow Reduction Facilities." This regulation defined "the greatest reasonable reduction" as *control of each CSO such that an average of one untreated discharge may occur per year.* Dischargers must achieve and maintain at least this level of control. In addition, treated and untreated overflows

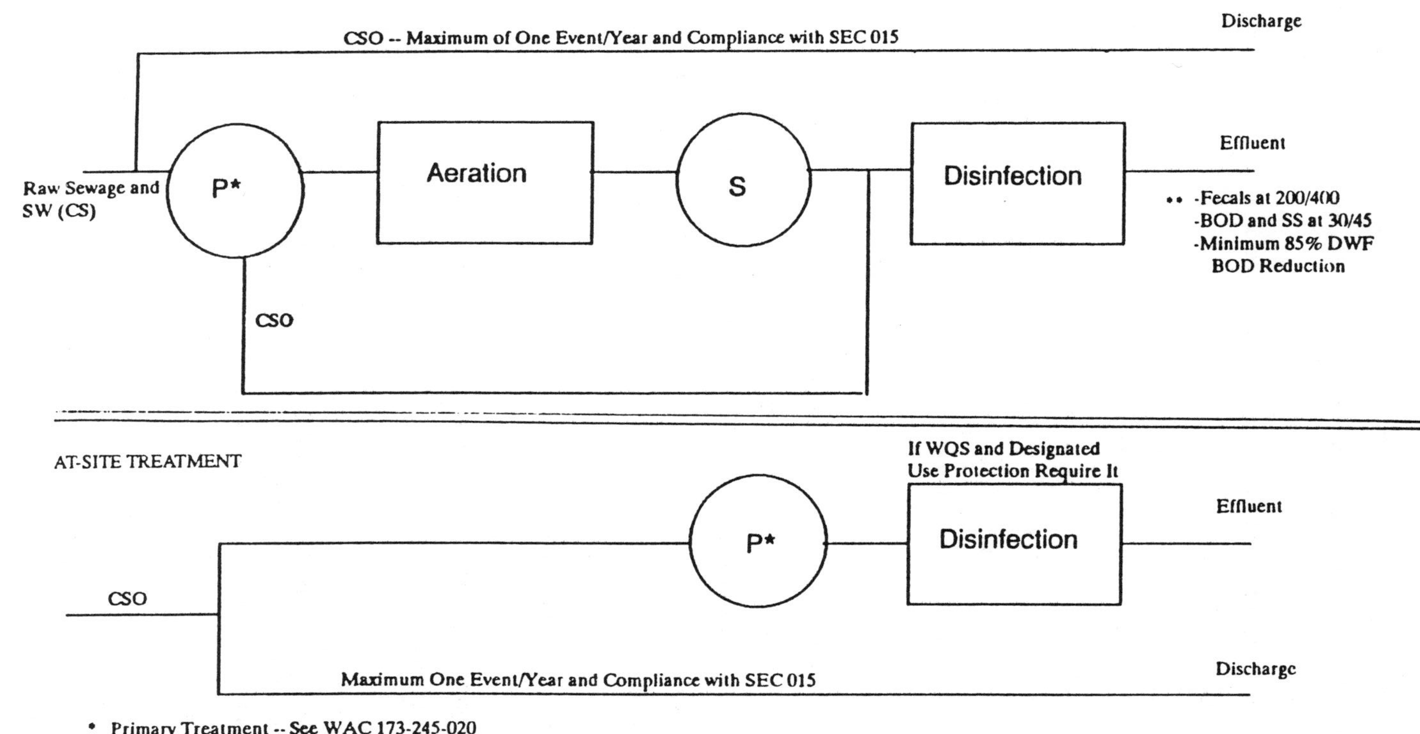

Figure 20. Schematic Representation of Ecology's CSO Treatment Requirements

must cause neither violations of applicable water quality standards nor restrictions to the characteristic uses of the receiving water, nor accumulation of deposits that exceed sediment quality standards. Therefore the regulation is both technology- and water quality-based.

The first required step is the submission of a CSO Reduction Plan. The plan must include the following:

Documentation of CSO Activity. This includes a field assessment and mathematical modeling study to establish each CSO's location, "baseline annual frequency and volume"; to characterize each discharge; and to estimate historical impact. Flow monitoring is almost always required. CSOs serving industrial/commercial basins require conventional, heavy metal, and organics analysis. The extent of sludge deposits must be confirmed with chemical analyses required for commercial/industrial areas.

A model must establish the rainfall stormwater runoff/CSO quantity relationship. A CSO baseline condition is set by using the model, the historical rainfall record, and the existing sewer system to estimate annual historical CSO volumes that would have occurred had the sewer system existed as it does today. A graph of annual estimated CSO volumes versus annual precipitation is developed. The baseline is taken as the 95 percent confidence limit line. It is that annual overflow volume that should not be exceeded 95 percent of the time given the corresponding annual rainfall amount.

Evaluation of Control/Treatment Alternatives. Municipalities must evaluate use of the following alternatives for controlling their CSOs:

- Delayed (storage) or direct transportation to the secondary treatment plant serving the sewer system. All flow reaching the secondary plant must receive at least primary treatment and disinfection with one bypass allowed per year. The plant must not violate its pollutant concentration effluent requirements. Storage and transport capabilities must be adequate to allow an average of only one untreated discharge per year at each CSO site.

- Onsite treatment equal to at least primary treatment, and offshore submerged discharge. Disinfection may be required for sites that are near or impact water supply intakes, shellfish beds, and primary contact recreation areas. The onsite system must be sized such that an average of one untreated discharge may occur per year. Primary treatment is defined as removal of at least 50 percent of total suspended solids and discharge of less than 0.3ml/l/hr of settleable solids.

- Separation of the combined system, creating distinct sanitary and stormwater systems.
- As interim measures to reduce pollutant loading, municipalities can explore use of best management practices, more restrictive sewer use ordinances, and pretreatment programs. Sewer maintenance programs to reduce infiltration and inflow are also acceptable control approaches.

Analysis of Proposed Alternatives. Proposed control alternatives must be analyzed for their water quality and sediment impacts. This includes the impact of discharges from new storm sewers. Construction and operation and maintenance costs must be estimated based on preliminary designs. Phased construction of control alternatives also should be considered.

Ranking and Scheduling of Projects. Proposed projects shall be given priority rankings using health, environmental (documented probable, and potential), and cost-effectiveness (e.g., cost per annual mass pollutant, volume, or frequency reduction) criteria. Municipalities shall propose compliance schedules based on these criteria:

- Total cost of compliance
- Economic capability of the community
- Other recent and concurrent expenditures for improving water quality
- Severity of existing and potential environmental and beneficial use impacts

Schedule Updates: Monitoring and Reporting. Annual reports must document frequency and volumes from each CSO location based upon field monitoring. If any CSO increases over its baseline condition, the municipality shall propose a project and schedule to reduce the CSO to or below the baseline condition. Annual reports must also explain the year's CSO reduction accomplishments.

Onsite treatment facilities shall have NPDES permits that limit effluent quality and quantity, and include reporting requirements.

CSO Reduction Plan amendments are due every 5 years in conjunction with the application for renewal of all applicable NPDES permits. The amendments shall assess the effectiveness of CSO reduction activities to date, reevaluate priorities, and propose new schedules based on current economic conditions and environmental knowledge.

Those activities and accomplishments scheduled to occur during the 5-year life of the municipality's NPDES permit shall be requirements of the permit or a companion administrative order.

Rationale for Selection of One Untreated Discharge per Year per Site, and Minimum of Primary Treatment.

- Interpretation of new CSO statute had to be consistent with historical interpretation of long-standing state water pollution control laws for technology-based treatment.

The state's water pollution control statues include a long-standing requirement that all wastes must receive "all known, available, and reasonable methods of treatment" prior to discharge to the state's waters, regardless of the quality of the receiving water, in order to prevent and control pollution. This is a technology-based law, somewhat equivalent to the federal Best Available Technology (BAT) economically achievable requirement. For each situation the state must decide what is technically possible and reasonable.

The historical interpretation of this requirement is that the term "reasonable" includes an economic consideration. In the context of municipal sewage treatment, the state used economic impact as a criterion for determining when secondary treatment was reasonable.

When the Water Pollution Control Statute was amended to include the requirement for the "greatest reasonable reduction of combined sewer overflows . . . at the earliest possible date," the Department of Ecology had to interpret that requirement in a manner consistent with the "all known, available, and reasonable" requirement.

- Interpretation of new CSO statute could not conflict with state water quality standards, nor the legislative record.

In addition, state law and regulation prohibit discharges that cause "pollution" or violate water quality standards. In reviewing the CSO discharge situations in the state, we determined that in almost every instance, an untreated CSO discharge was causing a water quality standard violation. This seems to require elimination or treatment of every CSO. However, the legislative record behind the new CSO statute seemed to indicate that the legislature intentionally stopped short of that requirement.

So, the Department of Ecology was faced with developing a technology-based CSO treatment/control standard that satisfied both the technology-based and water quality-based requirements of the law.

Minimum Treatment and Control Methods Identified. First, we reviewed the available CSO treatment technologies. Sewer separation and storage were long-standing, available control methods. Primary treatment of CSO had been applied in a number of states, including Washington. It was considered a minimum treatment level necessary to prevent violation of water quality standards outside a reasonable dilution zone. Technologies such as swirl concentrators would still result in the discharge and accumulation of sediments at the discharge locations. More advanced physical/chemical methods may be necessary to prevent violation of water quality standards in some situations, and should be required in those instances.

With minimum treatment and control methods identified, we still had two issues to resolve. Because CSOs are intermittent, and the list of reasonable technologies includes options for continuing untreated discharge, there had to be a determination of an allowable frequency for such untreated discharges. Secondly, the concept of economic reasonableness had to be factored in.

Maximum Allowable Frequency of Untreated Discharge Selected. All of the CSOs in Washington discharge to waters designated as Class A or AA. The characteristic uses of these waters include primary contact recreation, and in the case of marine waters, shellfishing. A review of past CSO control efforts within Washington and in other states led to the conclusion that one untreated overflow per year was the minimum necessary to protect all these receiving waters for their beneficial uses. In particular, CSO control planning efforts in the late 1970s and early 1980s by the Municipality of Metropolitan Seattle (METRO) and by the City of Seattle had concluded that this control level was necessary for the CSO receiving waters in the Seattle area in order to protect shellfishing and primary contact recreation.

The one untreated overflow per year level also seemed to satisfy the technology-based, "available" requirement. It had been achieved in a number of areas in the Seattle area. In addition, it was the strictest control level chosen in the San Francisco CSO control program. That program used a cost/benefit analysis to pick control levels ranging from one to 10 overflows per site per year. A control level of four overflows per year had been selected by the City of Sacramento.

One untreated overflow per year should also prevent significant recurring long-term water quality degradation. The one untreated overflow may still cause a temporary water quality standard violation. If monitoring confirmed such a situation, the state could require additional control.

"Reasonable" Economics Accommodated through Compliance Schedules. The final factor to consider was "reasonableness." Two years prior to the time of this decision, the Department of Ecology had decided to deny a number of municipal applications for a waiver from the requirement for secondary treatment. Such waivers were allowed under Section 301(h) of the Clean Water Act. The State Pollution Control Hearings Board and the Department of Ecology used economic impact as a criterion only for determining when such treatment would be required to be on-line.

In the context of CSO control, economic criteria were used to establish reasonable schedules for compliance with technology- and water quality-based treatment and control requirements.

National Cost for Combined Sewer Overflow Control

Atal Eralp, Norbert Huang, Michael Denicola, Robert Smith, and Tim Dwyer

Office of Wastewater Enforcement and Compliance
EPA Headquarters
Washington, DC

Background. There are approximately 1,050 communities nationwide with combined sewer systems with an estimated 10,770 combined sewer overflow (CSO) discharge points. These communities are primarily located in New England, the Middle Atlantic region, and the upper Midwest. Eleven states have 878 communities with combined sewer systems having approximately 8,700 CSO discharge points.

CSO Wastewater Characteristics. The primary components of CSO discharges are raw sanitary wastewater, industrial wastewater, and stormwater runoff. Characteristics of CSO discharge are dependent on the ratio of the three primary components. CSO discharges are characterized by the presence of fecal coliforms; total suspended solids; BOD; heavy metals (copper, lead, zinc, chromium); toxic organics (benzene, phenols, and other organic solvents); fertilizers; and pesticides. Other pollutants may be present in CSO discharges depending on the residential, commercial, and industrial profile of the system's service area.

CSO Wastewater Impacts. During heavy rains, as much as 90 percent of the total wastes that enter a combined sewer system never reach the POTW and are discharged untreated through CSO points into receiving waters. Some of the immediate effects of CSO discharges are:

- CSO discharges are wholly or partly responsible for the designation of 175 river miles and 4,400 lake acres in Kentucky and Tennessee as unsafe for recreation or fishing (Region 5 OIG Report, March 1990).

- Although only four Boston area communities have CSOs, they are responsible for the annual discharge of 9 billion gallons into Boston Harbor.

- CSO discharges resulted in the closing of beaches in the New York-New Jersey-Connecticut area (NY Times, September 5, 1989) and led to beach closings and shellfishing bans in Puget Sound (1987 Puget Sound WQM Plan, 1987), and the permanent closing of 25 percent and temporary closure of 10 percent of the Narragansett Bay's shellfishing beds.

Existing CSO Control Authorities. There are several existing nationwide CSO control authorities, including the Clean Water Act, the National Pollutant Discharge Elimination System (NPDES) permits, and EPA's August 1989 National Combined Sewer Overflow Control Strategy. The EPA strategy requires that states develop CSO permitting strategies that address the following elements:

- Inventory of communities with CSOs and the permitting status of their CSO discharges
- Prioritization of the unpermitted and inadequately permitted CSO discharges
- Issuance of single, system-wide permits
- Compliance schedules for technology-based permit limitations and applicable water quality standards
- Minimum technology-based limitations in permits and additional CSO control measures to meet the technology-based limits and water quality standards
- Monitoring of CSO discharges
- Adjusting water quality standards in limited cases to better address impacts in wet weather
- Funding and permit application forms

There are also several individual state CSO permitting strategies. To date, 30 states have submitted CSO permitting strategies. Twenty-one states have stated that they do not require CSO permitting strategies either because they have no communities with combined sewer systems or, if they have communities with combined systems, these communities have no CSO discharges. Of the 30 strategies that have been submitted, EPA Regions have unconditionally approved 19 and conditionally approved 3. Eight strategies are presently unapproved.

Key CSO Issues. Some of the key CSO issues that must be considered are technology-based standards, water quality standards, deadlines, and financing. Questions that relate to these issues include:

- Technology-Based Standards -
  - What minimum technology-based controls should be required of all communities with CSO discharges?
- Water Quality Standards -
  - Should CSO systems be required to immediately comply with applicable water quality standards regardless of costs?
  - Should EPA, as a matter of policy be encouraging states to downgrade current water quality standards to existing uses to reflect financial impacts?
- Deadlines -
  - Should statutory deadlines be extended for CSOs?
  - Should compliance with water quality standards be phased in over time?
- Financing -
  - What is the cost of CSO controls?
    - Current estimates: \$50 - \$200 billion
  - How should CSO programs be financed?
    - Current funding mechanisms: State Revolving Fund (SRF) Program, local funding sources
  - Should there be a federal role in future financing of CSO control programs beyond current Clean Water Act authorizations?
    - Maintain current approach (funding through SRFs)
    - Combination of SRFs and targeted grant funds to disadvantaged communities

<u>Alternative CSO Control Options</u>. There are several alternative options to control discharges from combined sewer overflows. The following section presents four options and their relationship to the key CSO issues.

Option 1: Implementation of EPA 1989 CSO Control Strategy.

Relationship to Key Issues

- Technology-based standards. Require six technology-based controls as minimum BCT/BAT, and consider other control measures where appropriate.
- Water quality standards. Require compliance with water quality standards. Evaluate whether appropriate to adjust water quality standards (downgrade, temporary variance, seasonal limits) to reflect economic impacts.
- Deadlines. Require cmpliance with existing CWA deadlines for both technology-based and water quality-based requirements. Since both these deadlines have passed, compliance schedules in administrative or enforcement orders must be used.
- Financing. References construction grants (which are no longer available) and SRF as available ways to obtain funding for CSO control program implementation.

Option 2: Legislative Action. Baucus/Mitchell S. 1081 requires states to inventory CSO discharges within 1 year, and cities to develop CSO Elimination Plans within 3 years of enactment. **CSO Elimination** is defined as no discharge during or following a 1-year/6-hour storm event. EPA must approve the municipal program within 3 months of submission only if 1) the statutory requirements for a "CSO Elimination Program" are met and 2) the municipality has adequate authority ***and financial resources are available.*** Cities must comply with these plans within 5 years for discharges to section 305(b)(2)(B) waters and 7 years for other discharges (2 year "good faith" compliance extensions are available).

Relationship to Key Issues

- Technology-based standards. Require the elimination of all CSO discharges during or following a 1-year/6-hour storm event. This in effect requires that any CSO discharge resulting from such a storm receive secondary treatment before discharge.
- Water quality standards. Require compliance with water quality standards.
- Deadlines. Require development of "CSO Elimination Programs" within 3 years and implementation of those programs within 7 to 9 years after EPA approval. (Note: Language provides that EPA cannot approve CSO Elimination Programs unless it finds a municipality has "adequate financial resources.")

- Financing. Provide grant funding authorization for states and municipalities up to a total funding level of $10,000,000 in FY92, FY93, and FY94. Extend eligibilities of other grants to include CSO controls. Extend eligibility of SRFs to include CSOs.

Option 3: Combined Sewer Overflow Control Act. This act was developed by the CSO Partnership. The proposal requires that:

- EPA must issue CSO permits in two phases.
- Phase 1 would require the elimination of dry weather overflows, proper operation and maintenance of the system to minimize wet weather overflows, maximize use of existing system, and implement the CSO study and plan requirement.
- Phase 2 must incorporate the technology-based and water quality-based controls.
- Requires compliance with water quality standards as soon as possible but specifies no firm deadline.
- Provides for the establishment of wet weather water quality standards and a variance from these standards when it is demonstrated that there is no reasonable relationship between the costs and the benefits of complying with these standards.
- Establishes a CSO Control Grant program ($500,000,000 in FY92 and in FY93). Grants can be used to fund control program costs beyond the limit of the city's financial capability to pay for the program with other funds which may be available.

Relationship to Key Issues

- Technology-based standards. Phase I permits require: 1) prohibition on dry weather overflows, 2) implementation of sewer system O&M practices to minimize CSOs, 3) maximum use of existing facilities to minimize CSOs, and 4) development of site-specific CSO study and control plans
- Water quality standards. Phase II permits require compliance with water quality standards "as expeditiously as practicable" taking into account the city's ability to pay *and* the availability of federal/state funding. *Cost/benefit variances* from water quality standards available where no reasonable relationship between the costs (economic and social) and the benefits of compliance.
- Deadlines. No specific date or schedule for compliance with permit conditions or achieving technology- or water quality-based requirements in the CWA. Phase II permit compliance is linked to financing and ability to pay.

- Financing. Establishes a CSO Control Grant program ($500,000,000 in FY92 and in FY93) for cities to use in implementing CSO control plans. Grants to be used to fund control program costs beyond the limit of the city's financial capability to pay.

Option 4: A policy paper was prepared by Gordon R. Garner, Executive Director, Louisville and Jefferson County Metropolitan Sewer District. It would require:

- Permittees must have a state/EPA approved I&M program within 1 year.
- Systems with significant water quality impacts from CSOs must within 5 years implement minimum requirements to eliminate floatables, eliminate excessive fecal coliform impacts, eliminate through pretreatment any IU discharge that causes toxicity violation of water quality standards, eliminate all CSO dry weather overflows as quickly as possible.
- CSO Abatement Plan must be developed when there are significant impacts from CSOs. Approvable CSO Abatement Plans include measures to reduce the CSO event frequency; the pollutant loading and volume of untreated discharges by 85 percent; the maximization of capture, storage, and secondary treatment of first flush flows at POTW if pollutant loadings in first flush are significantly higher than in later flows.
- States/EPA may require additional CSO control measures for CSOs discharging to selected receiving waters.
- Compliance deadline will be based on the cost of the abatement plan as a function of CSO population served or the percentage increase in wastewater rates needed to the finance abatement plan, whichever is longer.
- Abatement plans that cost more than 2,500 per equivalent customer or 1,000 per capita financed over 20 years are considered not to be cost-effective and can be scaled back. Permittees must still meet minimum requirements.

Relationship to Key Issues

- Technology-based standards. Systems with "significant water quality impacts" from CSOs must meet implementation requirements within 5 years.
- Water quality standards. Requires controls to avoid "significant violations" of water quality standards due to CSO discharges. Requires reduction in CSO impacts "at least" to the point that other impacts are more significant (point sources, nonpoint sources, naturally occurring background levels, etc.) and further improvements in water quality can be better achieved by using available resources to abate non-CSO impacts."

- <u>Deadlines.</u> Requires that minimum CSO control requirements be implemented within 5 years. Cities requiring a more stringent plan must include in the plan a schedule with appropriate interim target dates which become part of an enforceable compliance schedule. Generally, longer deadlines (up to 20 years) are established on a sliding scale linked to l) size of city, 2) total costs, or 3) percentage increase in wastewater rates.

- <u>Financing.</u> Does not address the financing of CSO control program implementation. Allows approved CSO Abatement Plans to be scaled back when they cost more than 2,500 per equivalent residential customer or 1,000/per capita financed over 20 years.

<u>CSO Control Technologies</u>. There are three types of control technologies to consider for combined sewer overflow systems. The following list presents typical examples of these CSO control technologies.

- *Nonstructural controls*

    - Street cleaning
    - Sewer flushing
    - Litter control
    - Industrial pretreatment programs

- *Minimal structure controls*

    - Proper operation and maintenance of regulators and overflow control devices
    - Screening of CSO outfall
    - Maximization of system's existing storage capacity

- *Structurally intensive controls*

    - Off-system collection and storage
    - Sewer separation
    - Elimination of infiltration and inflow
    - Swirl concentrators

<u>Cost Estimates</u>. As the experience with CSO control measures is very limited, to estimate the total cost for all CSO control measures is very difficult. Many assumptions are made in any attempt. Three approaches are presented here:

- Estimates based on needs surveys

- Estimates based on design storms
- Estimates based on experience with major cities

Estimates Based on Needs Surveys. Needs surveys basically estimate the needs based on four water quality goals: 1) aesthetics, 2) public health, 3) fish and wildlife, and 4) recreation. The following list shows CSO needs survey estimates for the last 12 years:

Year	Dollar Amount (1991 dollars) (in billion dollars)
1978	46
1980	57
1982	46
1984	15
1986	17
1988	20

The latest needs surveys include only 328 communities. The following is an attempt to extrapolate the needs survey total CSO cost to about 1,100 communities and allow some adjustments for recent up-to-date estimates:

328 communities	$20 billion
1,100 communities	$32 billion

Adjustment for up-to-date estimates $50 - $70 billion

Estimates Based on Capture and Treatment of a Design Storm. Total inches of rainfall varies greatly with frequency of storm. The following list of actual data for a location in New York demonstrates this variation:

Storage and Treatment (6 hour duration)

Design Storm	Total Inches
1 month	0.42
3 month	0.72
6 month	1.14
1 year	1.68
2 year	2.28
5 year	3.12

If a one year 6-hour storm is considered as the design standard, the total national cost can be estimated to be in the order of $150 to $200 billion. The assumptions for this estimate are as follows:

National Cost for 1-Year 6-Hour Storm Event

2.5 million acres of CSO area
55 billion gallons collected
$2.00 to $2.50/gallon storage
Add 40 percent for secondary treatment

$150 to $250 billion

The cost for different storms can also be estimated as the cost that will be approximately proportional to the inches of rainfall presented above. As can be seen, any change in the design storm will change the cost substantially.

Storage and Treatment Cost (6-hour duration)

Design Storm	Normalized Cost
1 month	0.25
3 month	0.43
6 month	0.68
1 year	1.00
2 year	1.36
5 year	1.86

Other Estimates. Several estimates were presented to the U.S. House of Representatives Subcommittee on water resources hearing on April 1991. These were based on the experience of the experts and varied from $56 to $322 billion.

Conclusions. Current plans for CSO control were briefly described. Several cost estimates based on these plans and other available data are presented. As the abatement programs are just starting and documented information is scarce, the estimates are based on many assumptions and estimators' own experience, and thus may vary greatly. Although the estimates vary from $30 to $300 billion, a range of from $50 to $150 billion appears most likely.

Stormwater Control for Puget Sound

Peter B. Birch

Washington Department of Ecology
Water Quality Program
Olympia, Washington

<u>The Puget Sound Water Quality Management Plan</u>. The 1991 Puget Sound Water Quality Management Plan requirements form the foundation of the stormwater program being developed by the Department of Ecology. The Plan was first adopted in 1987, updated in 1989, and again in 1990. The Plan and the Department of Ecology's stormwater programs apply to the cities and counties in the Puget Sound Basin.

<u>Local Stormwater Program</u>.

1) <u>Basic Stormwater Program for ALL Counties and Cities</u>

The Department of Ecology and the Puget Sound Water Quality Authority (PSWQA) will adopt rules to implement the local stormwater program. PSWQA's rule will emphasize procedures and processes while the Department of Ecology's rule will concentrate on standards and criteria. In addition, supplemental guidelines, including model ordinances, and a technical manual (see below) are being prepared to describe how local governments can implement their stormwater programs and meet the requirements of the rules.

The rules will set minimum standards for the following:

- Operation and maintenance programs are required for new and existing publicly owned stormwater systems.
- Runoff control ordinances will address drainage, clearing, and grading, erosion and sediment control, and protection of surface and ground water. They will apply to all new development and redevelopment.
- Local governments will be required to adopt ordinances to ensure maintenance and operation of privately owned stormwater facilities.

- Local governments will be required to keep records of new drainage systems and facilities.

Proposed schedule:

All cities and counties will be required to adopt ordinances and meet operation and maintenance requirements 18 months from the effective date of adoption of the rules.

2) Comprehensive Stormwater Programs for Urban Areas

Rules and guidelines will be prepared for comprehensive stormwater programs that will be implemented by all urbanized areas. This program is *in addition* to the requirements of the basic stormwater programs described above. Urbanized areas will be identified by the U.S. Bureau of Census definition.

These programs will address runoff from new and existing industrial, commercial, public facilities and residential areas, including streets and roads.

At a minimum, each urban stormwater program shall include:

- Identification of potentially significant pollutant sources and their relationship to the drainage system and water bodies.
- Investigations of problem storm drains, including sampling.
- A water quality response program, to investigate sources of pollutants, spills, fish kills, illegal hookups, dumping, and other water quality problems. These investigations should be used to support compliance/enforcement efforts.
- Assurance of adequate local funding for the stormwater program through surface water utilities, sewer charges, fees, or other revenue-generating sources.
- Local coordination arrangements such as interlocal agreements, joint programs, consistent standards, or regional boards or committees.
- A stormwater public education program aimed at residents, businesses, and industries in the urban area.

- Inspection, compliance, and enforcement measures.
- An implementation schedule.
- If, after implementation of the control measures listed above, there are still discharges that cause significant environmental problems, retrofitting of existing development and/or treatment of discharges from new and existing development may be required.

Proposed Schedule:

Six of the larger cities (Seattle, Tacoma, Everett, Bellevue, Bellingham, and Bremerton) and four early action areas will begin developing programs at date of adoption of the rules.

All urbanized areas will begin implementing programs by the year 2000.

Rule schedule:

The 1991 Puget Sound Water Quality Plan's target date to adopt the two rules is November 1991.

Technical Manual. The Department of Ecology will develop and update a technical manual for stormwater control practices. This manual will address:

- Erosion and sedimentation control at construction sites
- Detention/retention basins, infiltration, and conveyance systems
- Hydrologic analysis
- Control of pollution in runoff from urban land uses

The manual will serve as the minimum technical standards for local jurisdictions. Those that do not have their own manual may use the Department of Ecology manual; jurisdictions with their own manual must meet or exceed Department of Ecology standards.

Separate Technical Advisory Groups worked with the Department of Ecology to produce a draft of the manual that was released July 1990. This draft was reviewed and presently is being rewritten.

Proposed Schedule:

Another draft is being prepared for public review for July 1991.

Puget Sound Highway Runoff. The Department of Ecology has worked with the Washington State Department of Transportation (WSDOT) to adopt a rule and develop a program to control the quality of runoff from state highways in the Puget Sound basin. WSDOT will:

- Adopt a Highway Runoff Manual equivalent to the Department of Ecology's technical manual to enhance the quality of highway runoff
- Adopt a vegetation management program
- Include water quality best management practices (BMPs) as part of new construction projects
- Inventory and retrofit existing state highways with water quality BMPs where practicable
- Monitor where applicable
- Submit biennial reports

Public workshops were held on the draft rule in January 1990, at Pt. Townsend, Everett, and Tacoma, and meetings were held with the tribes. Public hearings were held in Bremerton on March 13, in Everett on March 14, and in Tacoma on March 15, 1991.

Proposed Schedule:

The highway program is scheduled to be adopted in May and be effective in June.

DISINFECTION

Total Residual Chlorine—Toxicological Effects and Fate in Freshwater Streams in New York State

Gary N. Neuderfer

New York State Department of Environmental Conservation
Avon, New York

Wastewater discharges frequently contain chlorine because it is used to disinfect potable water and wastewater effluents, to control biofouling in cooling water systems, and in industrial processes. The toxicity of chlorine to aquatic life has been the focus of extensive laboratory research, but few field studies have documented the effects of chlorinated discharges on freshwater aquatic life.

The fate of chlorine discharges to freshwater aquatic habitats has been the focus of considerable research. These studies have shown that dilution, phototransformation, chemical reaction demand and volatilization are the principle routes of chlorine dissipation in the aquatic environment. The importance of each fate route is variable due to chlorine species present in the effluent, physical characteristics of the receiving stream, and pH, temperature, turbidity and chlorine demand of the receiving water.

National Pollutant Discharge Elimination System (NPDES) and State Pollutant Discharge Elimination System (SPDES) permits for publicly owned sewage treatment plants (POTW's) often require chlorination of the effluent to control human pathogens, thus reducing the threat of waterborne infectious diseases. The U.S. Environmental Protection Agency (EPA) and New York State Department of Environmental Conservation (NYSDEC) have established ambient water quality standards (AWQS) for TRC to protect freshwater aquatic life. The EPA AWQS is 11 ug/L as a four-day average and 19 ug/L as a one-hour average. The NYSDEC standard is 5 ug/L for waters classified AA, A, B and C, and 19 ug/L for class D waters. These AWQSs were derived from toxicity data obtained from single-species laboratory toxicity tests with freshwater invertebrates and fish.

Project Objectives. The project scope was limited to the effects of TRC from POTWs that practice chlorine disinfection four or more months per year and discharge to lotic freshwater streams.

The study objectives were:

- To determine which POTWs were causing significant adverse effects to aquatic life
- To estimate the total stream distance in New York State that is affected by TRC from these discharges
- To use the study data to improve programs and polices on chlorine disinfection to minimize adverse effects on aquatic life

Study Methods. The dilution of effluents in receiving streams is easy to predict with reasonable accuracy. Plant design flows and receiving stream minimum average 7 consecutive day stream flow expected to occur every 10 years (MA7CD10) are determined as a routine part of the SPDES permit process. The factors affecting chlorine's fate in the aquatic environment have wide temporal and spatial variability. This study combined these factors by only looking at chlorinated effluent dilution and in-stream TRC concentrations. Effluent dilution rates at MA7CD10 flows were then correlated with in-stream TRC concentrations to predict which plants are likely to cause significant adverse aquatic ecosystem impact in the receiving stream.

Twenty-seven sites were studied during 1988 and 1989. The site-study techniques were designed:

- To identify significant TRC-caused impacts on aquatic organisms in the receiving stream
- To locate the TRC impact zone under varying physical conditions
- To predict the dilution rates where in-stream TRC would be < 5 ug/L

A battery of physical, chemical, and toxicological tests were used to evaluate each study site.

Tracer dye studies were used to determine in-stream time of travel, delineate the effluent zone of mixing, measure discharge and stream flows, and determine chlorine dissipation rate by comparing dye and TRC concentrations. Rhodamine WT dye was metered into the effluent near the outlet from the chlorine contact chamber. Fluorescence was determined with a portable fluorometer.

Chemical analyses included TRC, free chlorine, pH, ammonia, hardness, alkalinity, conductivity, dissolved oxygen, and temperature. Chlorine concentrations were measured using a modification of the amperometric titration method within 30 minutes of sample collection. The quantitation limit for this method was 5 ug/L chlorine.

Acute 48-hour aquatic toxicity tests were conducted on unchlorinated composite effluent samples and *in situ* (in receiving stream) upstream and downstream from the chlorinated discharge. *Daphnia magna* neonates (24 to 48 hours old) and fathead minnow larvae (4 to 10 days old) were used in both tests. The unchlorinated effluent tests defined non-chlorination induced effluent toxicity and helped isolate the TRC component of in situ toxicity.

All sites were studied during the low stream flow and warm water temperature period from June to September. Physical and chemical tests were done during daylight hours only, except at two sites where data were collected around the clock to determine diurnal trends.

Seven sites with year-around disinfection requirements were studied during the winter of 1988/89. These winter studies consisted of monitoring only the effluent and receiving stream TRC concentrations, without the other chemical, physical, or toxicological testing. The objective was to determine the relative extent of winter versus summer receiving stream plumes.

In-stream TRC Versus Dilution. Initial work with several analysis of variance and multiple regression statistical methods indicated that there was too much variability in the database to produce meaningful statistical results. That variability was not surprising, since dilution is only one of several factors that result in TRC dissipation in the receiving stream.

An empirical interpretation of the data base can be gained by looking at in-stream TRC based on the average concentration after mixing and at individual samples collected after mixing. The average TRC site concentration data indicate that the break-point for in-stream TRC concentrations exceeding the 5 ug/L AWQS lies somewhere between 32.9:1 and 39.5:1 dilution. At dilution ratios greater than this break-point, the receiving water is able to assimilate the effluent TRC. The individual water sample TRC data collected at the first downstream station where mixing was complete supports a similar conclusion. At dilutions greater than 40:1, only two samples had chlorine concentrations > 5 ug/L when the effluent TRC concentration did not exceed 2,000 ug/L. Both incidents were at a site where the stream temperature was less than 17°C or a chlorine slug may have occurred.

These empirical data interpretations were used to formulate a potential approach to SPDES permit criteria to assure compliance with the 5 ug/L AWQS. Plants with MA7CD10 dilution of greater than 40:1 were in the no-effect category and would receive a 2,000 ug/L permit limit. Plants with greater than 30:1 and less than or equal to 40:1 dilution would receive a permit limit of 1,000 ug/L. Those with greater than 15:1 and less than or equal to 30:1 dilution would receive a permit limit of 500 ug/L. All plants with MA7CD10 dilutions less than or equal to 15:1 would be required to use alternative disinfection, dechlorination or no disinfection.

This project was not designed to measure the relative importance of each major TRC dissipation route. But the data show that there was a significant and rapid initial loss of TRC other than dilution between the discharge and the point of complete mixing with the receiving stream. The rapid initial dissipation of TRC suggests that chemical demand was responsible for the major portion of this chlorine loss in the receiving stream. Other studies have documented this initial chlorine loss as well. In this study, the mean initial loss was 70.2 percent (SD = 19.2 percent). The majority of the study sites were located on small, heavily shaded, and low-gradient streams where mixing was complete within a few minutes. Most of the effluent TRC was combined chloramine species. Because chloramines have been shown to be less subject to photodegradation than free chlorine, it is unlikely that photodegradation played a major role in chlorine dissipation. The non-turbulent nature of the study streams was likely to minimize volatilization losses as well.

Kilometers of Stream Affected. An estimated total of 158.4 kilometers of stream have mean TRC concentrations that would exceed the 5 ug/L AWQS at MA7CD10 flow during the summer. Compared to a total of 129,710 kilometers of freshwater streams in New York State, those potentially adversely affected by TRC from POTW's comprise 0.12 percent of the total. Had winter data and more nighttime data been included in the estimate, this would have increased significantly.

Diel In-Stream TRC Concentrations. Effluent and in-stream TRC concentrations were analyzed on a 24-hour-a-day basis at two sites. Effluent and in-stream TRC concentrations were significantly higher ($p=0.05$) during nighttime than during daylight hours.

Photodegradation of TRC has been documented in receiving streams. An EPA study found that when chlorine alone was added to artificial streams (assumed TRC mostly present as free chlorine), there was a strong diurnal trend in in-stream TRC concentrations due to photodegradation during daylight hours. When chlorine and ammonia were added to closely simulate a wastewater treatment plant discharge (assumed TRC mostly present as combined chlorine), no in-stream diurnal trend was observed. The TRC at the two sites in this study was present as combined chlorine, yet there were strong diurnal trends in in-stream TRC concentrations.

The diurnal TRC trends were due to increased mass loading of TRC from the effluent to the receiving stream at night. Effluent mass loading is defined as effluent TRC concentration times effluent flow. Mass effluent TRC loading was proportional to in-stream TRC concentrations.

***In situ* TRC Toxicity to *D. magna* and Fathead Minnow.** This project was not designed to produce an LC50 for *D. magna* neonates and fathead minnow larvae under field exposure conditions to TRC. The *in situ* toxicity test data from various sites was composited to get an acute dose-response curve. There were indications in the data base that the toxicity of free and combined chlorine toxicities might be different. Only data from sites where the majority of effluent TRC were combined chlorine were used in the LC50 calculations. Data from sites with upstream toxicity were also eliminated. Depending on which data points were used and the

LC50 calculation method, the 48-hour LC50s ranged from 4.1 to 6.1 ug/L TRC for *D. magna* neonates and 30 to 70 ug/L for fathead minnow larvae. The laboratory-derived species mean acute values for these two species are 28 and 106 ug/L TRC, respectively. These field data LC50s indicate that in-stream TRC toxicity is higher than in the laboratory.

There are several factors that might affect the field LC50s. These factors include:

- In-stream TRC concentrations were higher at night, but most of the data used in the LC50 calculation were from the daytime
- Considerable variation in the exposure concentration
- May be difference in chlorine species tested between laboratory and field studies
- Toxicants in effluent and receiving stream
- Physical stresses of *in situ* exposure

Catfish (*Ictalurus punctatus*) have shown a similar response under field exposure conditions. The catfish species mean acute value is 90 ug/L. When channel catfish were exposed in streams where only chlorine was added, so the majority of TRC was presumed to be present as free chlorine, there was no significant mortality at TRC concentrations as high as 183 ug/L. When chlorine and ammonia were added, so it was assumed that most of the TRC was present as combined chlorine, there was complete channel catfish mortality at 25 ug/l and reduced growth at less than 1 ug/L TRC. There is a clear indication that under field exposure conditions combined TRC is significantly more toxic than expected and more toxic than free TRC.

Fate of Free Versus Combined Chlorine. It is generally accepted that combined chlorine degrades slower than free chlorine. An EPA study found that in-stream TRC was more persistent when only chlorine was added to a study stream. The data from this study indicated that after initial chemical demand by the receiving water, free chlorine is more persistent than combined chlorine. It is possible that the free chlorine analysis method is measuring a more persistent chlorinated organic chemical formed during the chlorination process. This unknown chlorinated organic compound reads as free chlorine in the test, but it is more persistent and less toxic to aquatic life than combined chlorine species. This solution is only speculative, and there could be other explanations for these observations.

<u>Summer Versus Winter TRC</u>. The data indicate that the average effluent TRC concentrations were higher and in-stream plumes extend farther downstream during the winter. This was likely due to reduced rates of chemical demand reactions, photodegradation, and volatilization at colder water temperatures.

The impact of larger in-stream TRC plumes on aquatic life are unknown. The literature indicates that at colder water temperatures the toxicity or rate of toxicity to aquatic life is reduced. Chlorine toxicity at cold stream temperatures has not been well characterized.

The data from this study indicate that toxicity or rate of toxicity decrease proportional to stream temperature. The lack of mortality to *in situ* test organisms at some sites appeared to be due to reduced stream temperatures.

<u>Compliance with SPDES TRC Permit Limits</u>. The data base identifies a serious problem of inadequate control of effluent TRC concentrations to protect aquatic life in the receiving stream while achieving adequate effluent disinfection. During the summer studies, 14 percent of the effluent samples analyzed exceeded the SPDES permit limits for TRC. This increased to 44 percent of the samples during the winter sampling. Out of 27 sites, only 10 were always in compliance with their SPDES TRC permit limits. Two of these plants had electro-chemical feedback systems, one was flow-proportional, and the remainder had static chlorinators.

Sites with static chlorinators, with or without manual diel adjustment of the chlorine feed rate, in general were not able to consistently meet their SPDES TRC permit limit. Sites with electro-chemical feedback systems that analyze the effluent TRC concentration and automatically adjust the chlorine feed rate did the best job of controlling effluent TRC concentrations. This was especially true if the systems were maintained daily.

EPA Disinfection Policy and Guidance Update

Robert Bastian

U.S. Environmental Protection Agency
Washington, DC

State/EPA Task Force. A joint state/EPA Task Force was formed in 1988 to review EPA's position on municipal wastewater disinfection. This review of the Agency's disinfection policy was initiated to address concerns about the potential adverse effects on aquatic life and human health from wastewater chlorination. In part this issue was raised by an EPA study released in 1986 that suggested that up to two-thirds of the POTWs were likely to discharge wastewater that exceeds the acute freshwater chlorine criteria, putting some 3,500 different water bodies at risk. The Task Force reviewed information which became available since 1976 to determine if changes were needed to the existing policy (issued in 1976), which calls for disinfection requirements to be set on a case-by-case basis, consistent with applicable state water quality standards for bacterial indicator organisms and for chlorine.

A 2-day workshop was held in November 1988 to help summarize the current status of information associated with a variety of topics relevant to disinfection of municipal wastewater discharges including: the status of alternative disinfection technologies; water quality criteria for chlorine and indicator organisms; and case history studies, including infield studies of residual chlorine toxicity and application of alternative indicator organisms. The workshop and other sources of information served as the basis of a technical support document and proposed draft language designed to strengthen the existing case-by-case policy.

Task Force Findings and Conclusions. In general, the Task Force found that while wastewater disinfection is necessary to protect public health, as currently practiced it may present significant risks to aquatic life. These risks can be lessened by reducing disinfection where unnecessary or excessive and utilizing dechlorination or alternative technologies to chlorination. In addition, the Task Force agreed that the risks to aquatic life due to disinfection, and alternatively, the risks to public health from reducing disinfection, are not fully understood.

Some of the specific conclusions reached by members of the Task Force based upon their examination of disinfection policies, practices, and the existing scientific data included the following:

- Residual chlorine is toxic to aquatic life at the low levels produced in wastewater treatment effluents, but more information is needed on the actual instream risks, particularly chronic effects, in order to better understand the impacts that chlorine and other disinfectants have on aquatic life.

- Although chlorination produces organic chlorinated by-products in wastewater, some of which are toxic and potentially carcinogenic, they are generally not considered to be a significant human health concern in wastewater effluents because they are produced at low levels. However, not enough is known about the possible aquatic life effects of these compounds, such as bioaccumulation and chronic toxicity.

- Numeric chronic water quality standards developed by the states, based on EPA criteria or site-specific criteria, generally protect aquatic life from adverse impacts due to unacceptable levels of chlorine toxicity.

- Reduction of wastewater disinfection with chlorine will protect aquatic life by reducing the levels of chlorine to which aquatic organisms are exposed. This can be accomplished through seasonal disinfection, lower levels of disinfection, use of alternative disinfection practices, and elimination of disinfection where appropriate. More information is needed on the relative public health risks and aquatic life benefits of changing disinfection practices.

- Improvements in the efficiency of chlorination can effectively reduce chlorine discharges at many treatment plants. Where further reductions are needed, technological modifications to the treatment process will be necessary.

- Dechlorination is generally easily and economically retrofitted to chlorination facilities. However, dechlorination may not remove all potential toxicity. Concerns have been raised that even an infrequent failure of a dechlorination system that allows chlorine to enter receiving waters could substantially impact aquatic life. Alternative disinfection technologies such as ultraviolet radiation and ozone are often less hazardous to aquatic life than chlorination and are becoming viable both technologically and economically.

- New indicator organisms such as enterococci and *E. coli* can more accurately determine risks associated with contaminated water than can fecal coliform measures. However, further research may be needed to determine the appropriateness of these indicators in certain situations.

Proposed Revised Policy Language. Proposed draft policy language, which was widely circulated for comment, emphasized reducing or eliminating wastewater disinfection requirements where possible to do so without adversely affecting public health. It also stressed that the protection of aquatic life from the adverse impacts of chlorination residuals and by-products may require operational modifications at wastewater treatment facilities to improve the efficiency of chlorination practices, dechlorination following chlorination, or the use of alternative forms of disinfection. Comments received from the Task Force members, state water resources agencies and health departments, and other interested parties in response to the proposed draft language were mixed, some supporting the stronger language and others strongly opposed to any "weakening" of disinfection requirements for fear of increased public health impacts. Many reviewers noted that the draft did not functionally change the existing case-by-case policy, and noted the need for better guidance to assist states with evaluating the potential public health and aquatic life impacts of reducing or eliminating disinfection requirements.

Results of Task Force Policy Review. The results of this policy review effort have lead to a decision by EPA *not* to issue a new or revised policy statement, but to restate and emphasize the existing case-by-case policy issued in 1976 and to provide updated technical guidance. The outputs generated by the review of EPA's position on municipal wastewater disinfection include the following:

- A foldout titled "**Municipal Wastewater Disinfection: Protecting Aquatic Life and Human Health from the Impacts of Chlorination**" (dated February 1991) to be widely circulated
- An updated version of the proceedings of the municipal wastewater policy review Task Force workshop in the form of a book edited by and with an overview prepared by Dr. Charles Noss to be published by Lewis Publishers in 1991
- A "**Municipal Wastewater Disinfection State-of-the-Art Document**" produced from the policy update technical support document to be published by EPA during 1991
- A brochure summarizing the State-of-the-Art document to be published by EPA during 1991

In addition, as funding allows, a methodology with sample case studies is being developed to help states with evaluating the potential public health and aquatic life impacts of reducing or eliminating disinfection requirements. Finally, consideration is being given to updating the Agency's 1974 shellfish sanitation guidance document "Protection of Shellfish Waters" to better reflect the current state-of-the-art and explore potential improvements to current techniques for protecting shellfish sanitation.

CONSTRUCTED WETLANDS

Use of Constructed Wetlands to Treat Domestic Wastewater, City of Arcata, California

Robert A. Gearheart

Environmental Resources Engineering Department
Humboldt State University
Arcata, California

<u>Introduction</u>. In recent years, there has been an increasing need for the development of improved cost effective methods for wastewater treatment, specifically for those communities which could be categorized as small to medium in size. While the "progress" of our industrial society continues at a rapid rate, technological advances in treatment methods for the new variety of toxic chemicals, exotic organics, and general domestic sewage seems stymied. Initial construction cost and continuing operational costs of wastewater treatment plants are the most significant factors affecting the technology selection process. Along with the consulting engineer's lack of understanding of natural process, this has focused the need to consider alternative and innovative wastewater treatment processes. The cost to small communities for reaching the same level of wastewater treatment as large communities using standard technology is disproportionately high. Although large sums of money have been made available by the Federal and State governments for pollution control systems, relatively few funds are being applied to advance research and development of improved treatment technology. Since present wastewater treatment systems are primarily designed after "natural" mechanisms for pollution abatement (trickling filters, activated sludge, oxidation ponds, etc.), it is ironic that reliable, cost-effective, and efficient treatment of wastewater utilizing controlled nutrient uptake by macrophytes and microbial communities in a marsh is not in wider use and encouraged by regulatory and funding agencies.

Use of wetland wastewater treatment systems based on emergent plant species and their associated microbial communities is more widespread than use of floating aquatic plant systems. Most wetland processes involve the growth of rooted emergent plants such as reeds and bulrushes in an artificial bed and the passage of wastewater either across the surface of the wetland (surface-flow systems), or through the growing medium in which the wetland plants are rooted (subsurface-flow or root zone systems).

Surface-Flow Wetlands—General. The surface-flow wetland approach utilizes the stems of wetland plants as the main site for effluent treatment. In this method, beds of emergent wetland plants, such as reeds or bulrushes, are flooded with pretreated effluent which is retained within the wetland system for a predetermined period prior to discharge.

Surface-flow wetland plant stems provide a substratum for the microorganisms which achieve the desired effluent treatment. Wetland processes result in an accumulation of organic material in the bottom of the system where microorganisms also occur in high densities and further enhance effluent treatment, particularly in terms of nitrogen elimination (Bartlett et al., 1979) and anaerobic decomposition of detrital material to carbon dioxide and organic acids. Figure 21 depicts the processes involved as suspended solids are removed in the initial volumes of a wetland treatment system. The City of Arcata, California, has been experimenting for 8 years with twelve 6 x 10 meter pilot project cells, two 2½ acre wetland treatment cells, and 31 acres of effluent receiving surface flow wetlands (Gearheart, 1985).

Subsurface Flow Wetlands—General. The principle behind the subsurface-flow wetland treatment system involves passage of wastewater through a specially prepared soil, sand, or gravel medium in which reeds or other emergent plants are grown. Wastewater treatment occurs in the growing medium, principally as a consequence of the growth of wetland plant rhizomes, which are claimed to enhance the hydraulic conductivity of the growth medium and introduce oxygen into adjacent areas of the growing medium.

Surface Flow Wetlands—Natural. Discharge of pretreated wastewater to natural wetlands has been a widespread practice for many years in the United States, where a number of sites have been identified as being the subject of ongoing discharge for more than 50 years (Hammer and Kadlec, 1983). Discharge sites have been reported from Florida to Canada's Northwest Territories (Nichols, 1983). In the United Kingdom, wetlands have been used for treatment of wastewater for more than 100 years at some sites (Cooper and Boon, 1987).

In general, few water quality problems have been observed with discharges to natural wetlands, but the assimilative capacity of natural wetlands has only been monitored in detail since 1960 (Knight, 1985), and in the light of insufficient long-term data some workers advocate caution in the management of natural wetlands to which wastewater is discharged (Nelson and Weller, 1985).

Constructed Wetlands—Specific. The methodology of constructed wetlands, based on constructed basins planted with wetland species, was pioneered by Dr. K. Seidel in the 1950s (Rossiter and Crawford, 1984).

This methodology is presently utilized in the Netherlands, where it is applied to intermittent flows such as campsites, or sewage disposal from small communities (DeJong and Koridon, 1985).

Use of constructed wetlands for surface-flow effluent treatment in the United States and Canada is presently restricted to a series of pilot scale trials at Arcata in northern California (Gearheart et al., 1982; 1983; 1985), Gustine in southern California (Crites and Mingee, 1987), Orlando and Lakeland in Florida (Feeney et al., 1986), Listowel (Herskowitz et al., 1987), and Port Perry in Ontario, Canada.

It is emphasized that wetland treatment results might be interpreted in relation to the particular characteristics of the wetland system for which the results have been derived. Accordingly, treatment data presented in this section should be interpreted in the light of associated data on wetland characteristics. The following results generally represent the level of treatment effectiveness reported for such systems and are not intended to comprise a definitive compilation of wetland treatment capabilities.

BOD. BOD (biochemical oxygen demand) is a measure of the oxygen uptake in a given aquatic system principally as a result of the biochemical processes of the microorganisms in that system. High levels of BOD in wastewater can result in dissolved oxygen depletion of the receiving waters to which wastewater is discharged. The City of Arcata's pilot project showed that lower hydraulic loading rates produced higher BOD removals. Seasonal variations in effluent concentration were affected by vegetation type, density, and distribution. The rate varied from 41 to 65 percent. Those cells loaded at lower rates consistently produced BODs of 20 mg/L or less. Retention periods of 200-300 hours produced BODs of 10 mg/L or less.

The lowest BOD treatment efficiency value was reported for a subsurface-flow wetland which was reported as not being operated effectively (Lienard, 1986). For remaining sites which were operating correctly, reported BOD reductions ranged from 97 percent reported for long-term surface-flow wetlands in the Netherlands (Greiner and DeJong, 1982), to 56 percent reduction reported as an overall average for a surface-flow system at Arcata (Gearheart et al., 1983). At Arcata quarterly BOD reductions ranged from 25 percent to 88 percent for oxidation pond effluent, and reduction was found to be significantly influenced by temperature and low rate.

Constructed wetlands were able to process shock organic loads with little to no effect on effluent quality. Figure 22 shows a range of organic loadings from 30 lbs/acre/day to 300 lbs/acre/day (Gearheart et al., 1985). The effluent BOD did not significantly increase until organic loadings of 200 lbs/acre/day or greater were observed. Suspended solids levels, on the other hand, were minimally affected even at this higher loading. There are very few wastewater treatment processes that can produce an acceptable effluent quality over an order of magnitude increase in the BOD and suspended loading to the system. This ability to accept shock loads and to recover without any external process control is an important characteristic of wetland treatment systems. Figure 22 shows these relations as the effluent BOD remains low and stable up to the 150 lbs BOD/acre/day loading. At these higher loadings, though, the pounds of BOD removed are a function of the loading. This suggests that a wetland system can serve both as a roughing and a polishing system in a wastewater treatment train as it applies to BOD and suspended solids removal.

The wetland system affords a complex microbial community which processes both particulate and dissolved organic material as it moves through the various communities. The effect these microbial communities have on BOD removal can be seen in Figure 23.

It is clear that properly established and operated wetland treatment systems have the potential to significantly reduce BOD levels in wastewater, but that wastewater influent characteristics and wetland design will have a major influence on the level of BOD reduction.

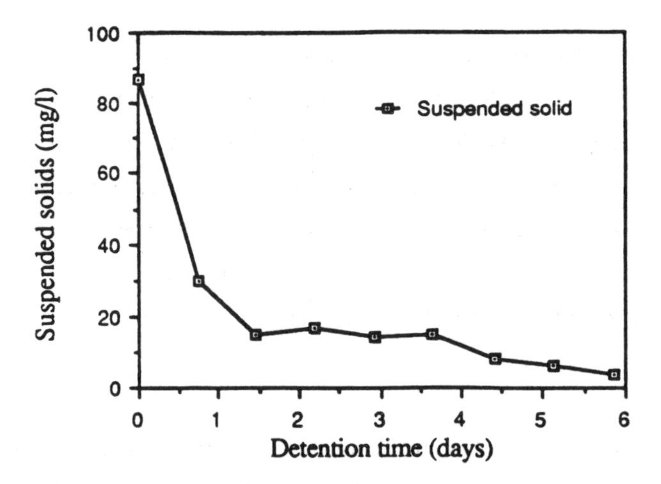

Figure 21. Suspended Solids Removal as a Function of Organic Loading over a 55 Week Period

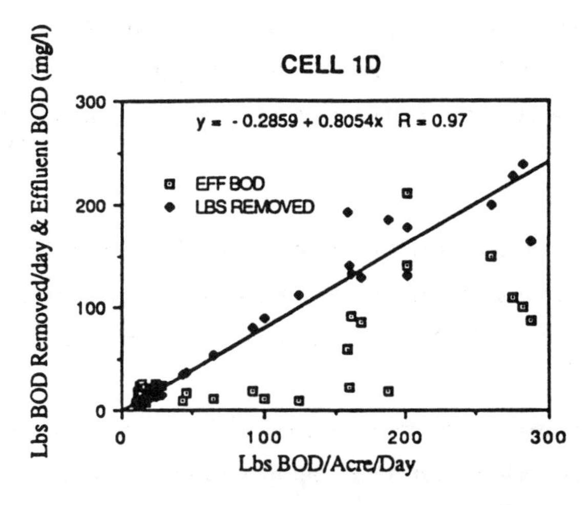

Figure 22. Regression Curve of BOD Removal versus BOD Loading to Arcata Pilot Project

Suspended Solids. The suspended solids (nonfilterable residue) content of wastewater is of direct water quality significance in terms of turbidity in receiving waters, and indirectly in relation to the associated transport of other waste constituents such as nitrogen, phosphorus, and BOD.

Suspended solids were removed in the first sections of the Arcata pilot project wetlands. This represents a theoretical retention time of about one day. This autoflocculation/sedimentation process builds a significant detrital bank in the upstream section of the wetland. This detrital bank is about 70 to 90 percent of the volume in this first section after 8 years of continuous loading. The detrital bank extends in a tapered fashion 75 percent of the length of the cell (Gearheart et al., 1985). Organic loadings of suspended solids are approximately zero order kinetics over the range 0 to 200 kg/ha/day, reflecting this progressive accumulation through the cell length (Figure 24).

The effectiveness of constructed wetlands to treat domestic effluent can best be seen, as in Table 8, by comparing the 8 years of research and monitoring in Arcata (Gearheart et al., 1982, 1985). Table 8 shows the removal efficiency and effluent quality of the two pilot projects and full-scale AMWS. The variations in suspended solids can be attributed to the high fraction of open water at the Arcata Marsh and Wildlife Sanctuary compared with the pilot project's densely vegetated water volume. The effectiveness of wetland systems to consistently (8 years of data to date) remove SS and BOD at a level significantly below secondary standards is noteworthy.

The dissolved oxygen levels in the wetlands is a function of the organic loading and the fraction of open water. In the first pilot project, the cells were totally vegetated by the end of the 2-year study. The average dissolved oxygen level was lowest in this study at 1.1 mg/L. Compare this with the full-scale project where the open water fraction is 75 to 90 percent and where the average dissolved oxygen was 5.0 mg/L reflecting the oxygen input from phytoplankton populations.

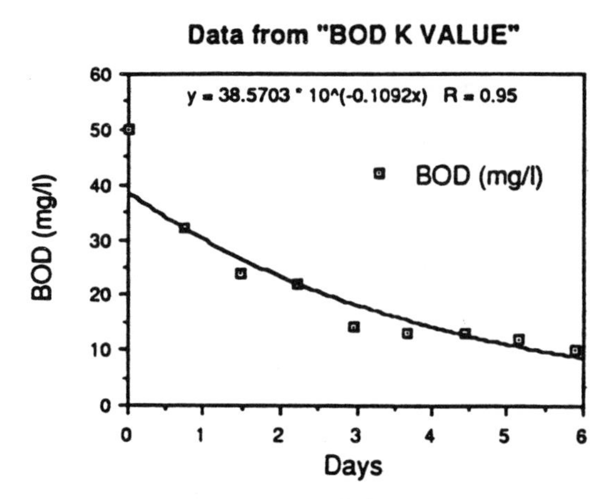

Figure 23. BOD Removal through Pilot Project Cell Showing the First Order Removal of BOD through a Compartment Cell

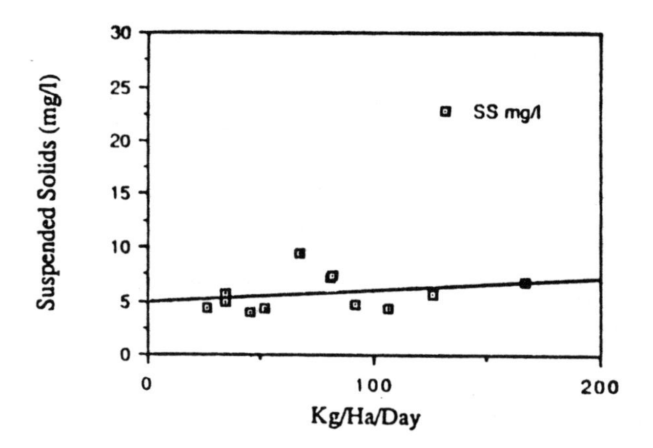

Figure 24. Suspended Solids Removal through Pilot Marsh 8 (Samples Taken at 7 Weirs through the 200-ft length)

Nitrogen and Phosphorus. The nitrogen component of wastewater is of water-quality significance (along with phosphorus) in relation to the potential enrichment of receiving waters, which can lead to excessive algal growth and eutrophication.

The rate of nitrification is dependent on temperature and the oxygen availability in the wetland (Stowell et al., 1981), and the process is only possible where oxygen can readily diffuse to the reaction site (Hammer and Kadlec, 1983). For this reason, anaerobic wetlands which may have been subject to high BOD loadings and surface flow wetlands which for a number of reasons may have low dissolved oxygen levels, will not be effective as nitrification systems—for example, Arcata (Gearheart et al., 1983). Oxygen translocation by plant roots has been reported as potentially useful in this regard, particularly if the wetland plants are grown hydroponically (Stowell et al., 1981).

In general, it appears that surface-flow wetlands are not effective nitrifiers as a consequence of low dissolved oxygen levels, but they are potentially effective denitrifying systems in view of the presence of anaerobic areas. Surface-flow wetlands would therefore potentially be very effective in nitrogen removal for highly nitrified effluents.

Subsurface flow systems have been found to be relatively poor at denitrification unless supplemental carbon is added (Gersberg et al., 1984, 1986). However, in view of the potential for oxygen translocation by the roots, subsurface flow systems are potentially valuable in nitrification.

Reported nitrogen removal efficiencies for wetlands vary for surface flow wetlands from around 26 percent as an average at Arcata (Gearheart et al., 1983), to 88 percent for a long-term detention system in the Netherlands (Greiner and DeJong, 1982); and for subsurface-flow systems 13 percent for an inefficient system at Kalo in Denmark (Brix and Schierup, 1987) to 95 percent for a carbon supplemented system at San Diego (Gersberg et al., 1984; 1986).

The ammonia nitrogen levels from the oxidation varies significantly as a function of the algae and bacteria. As the phytoplankton population grows, ammonia is taken up by the plants. As zooplankton reduce the population and excrete ammonia as a byproduct, the levels go back

Table 8. Comparison of Pilot Project and Full-Scale Results for NPDES Parameters (Percent Change Oxidation Pond Effluent through Wetlands) 1980-1988

	First Pilot Project, All Cells	Second Pilot Project, All Cells	Full-Scale Operation
BOD			
Mean	11.4	13.8	12
% Change	-56	-73	-55
% Less 30 mg/L	100	100	100
% Less 20 mg/L	84	72	81
% Less 10 mg/L	37	33	18
SS			
Mean	5.3	10.8	14
% Change	-85	-80	-54
% Less 30 mg/L	100	100	100
% Less 20 mg/L	100	93	78
% Less 10 mg/L	91	43	45
Dissolved Oxygen			
Mean	1.5	1.1	5.0
% Change	-73	-76	-27
pH			
Mean	6.5	6.1	7.1
% Change	29	-14	-6
Theoretical Retention Time/s (days)	1.5-30		
1981 Average	3.7		
1982 Average	9.0		
Open Water Fraction	0-10	0-25	75-90

up again. Very little ammonia is oxidized to nitrates in the oxidation ponds. The nitrifying bacteria need attachment sites which are not found in the open water volumes of an oxidation pond. As can be seen in Figure 25 (Gearheart et al., 1985), the influent varies from 3-4 mg/L to 35 mg/L over the study period of March through July.

Figure 26 shows the same time period with two different loading rates. Cell 5 was at 2.94 gpd/ft^2 and cell 11 was loaded at 1.47 gpd/ft^2. A similar phenomenon can be seen in Figure 26 compared with Figure 25. The only difference is that the vegetative system became saturated faster for the higher loading rates. Cell 5, for example, did not remove ammonia nitrogen to the 1 mg/L level. Cell 11 which was loaded at 3 times the rate of cell 3 saturated at the 18th week, while cell 1 saturated at the 25th week.

It is concluded that nitrogen removal by surface-flow or subsurface-flow wetlands is presently relatively consistent as a function of temperature, plant density, and nitrogen loading. However, wetlands have a number of important attributes which should lead to effective nitrogen removal, including wetland soil-water characteristics and an inviting environment for denitrifying bacteria. Work with carbon supplementation (Gersberg et al., 1984; 1986) indicates that nitrogen removal mechanisms can be optimized and that the means of optimizing these nitrogen removal mechanisms is clearly an area for active research (Howard-Williams, 1985).

The removal of phosphorus from wetland systems is intermittent, and little is understood about the mechanisms involved in uptake. The principal phosphorus removal mechanisms are precipitation and adsorption to sediments, with secondary mechanisms including plant uptake and sedimentation (Tchobanoglous, 1987). Phosphorus is rapidly immobilized in organic soils, and thus saturation is reached reactively rapidly with the process being partially reversible (Hammer and Kadlec, 1983). Ultimate removal of phosphorus from wetland systems could be achieved by harvesting of plants, dredging of sediments, or resolubilizing of phosphorus stored in sediments and released to receiving waters when it would have the least environmental impact (Stowell et al., 1981).

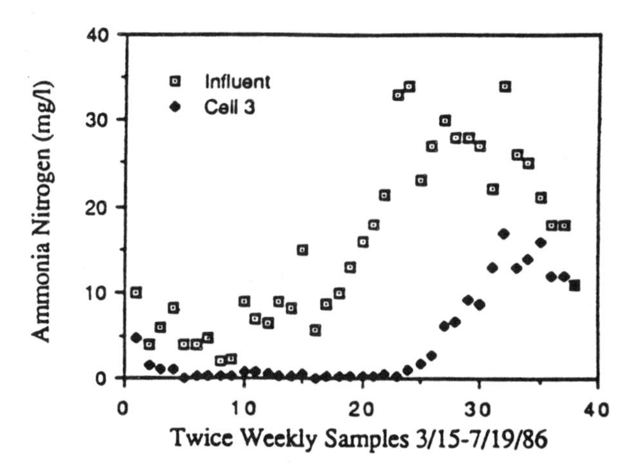

Figure 25. Ammonia Nitrogen Levels in Arcata Pilot Project Influent and in the Effluent from Cell 3

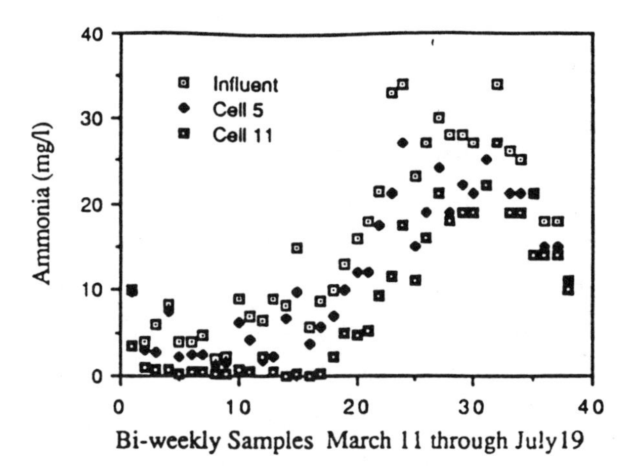

Figure 26. Ammonia Nitrogen Levels in Arcata Influent and in Effluent for Cells 5 and 11

Phosphorus is commonly released during winter in wetland systems (Wile et al., 1981; Stowell et al., 1981) and this would be consistent with release during time of minimal environmental sensitivity.

Phosphorus uptake by the macrophytes can be seen from data collected at the Arcata pilot project study (Figure 27, Gearheart et al., 1983). The removal of soluble phosphorus from the water column, approximately 3 mg/L, is correlated with the growing season for a hydraulic loading rate of 0.5 gpd/ft^2. This represents about 30 acres per million gallons of secondary effluent. Uptake occurred for only the four months of the growing season. Over the whole year, the phosphorus removal was approximately 10 percent at the lower loading rates and 0 percent at the higher loading rates.

Reported wetland removal results indicate a variable wetland performance with net phosphorus removal rates ranging from 0 percent in subsurface-flow systems (Phillips et al., 1987) and surface-flow systems (Gearheart et al., 1983) to 79 percent for a long-term surface-flow system in the Netherlands (Greiner and DeJong, 1982) and 83 percent in sand/soil based subsurface-flow systems in Denmark (Brix and Schierup, 1987).

Metals. In wetlands, metals are removed from wastewater by plant uptake, chemical precipitation, and ion exchange with and adsorption to settled clay and inorganic compounds. However, it is likely that the potential capacity of wetlands to remove metals by plant uptake and harvesting will be small, and ultimate removal of metals from wetlands systems will probably be most effectively achieved by methods for the removal of phosphorus (Stowell et al., 1981).

Fecal Coliform Removal. Wetlands have been shown to effectively remove fecal coliform organisms (Gearheart et al., 1985; Gersberg et al., 1984). The mechanisms for removal have been suggested to be flocculation sedimentation, adsorption, temperature ingestion, and denaturing (OR potential, UV light, etc.) (Ives, 1986). Data from the Arcata Marsh pilot project show a significant removal of fecal coliform in an experimental cell (20 x 200 ft) with a theoretical detention time of 6 days. The removal data fit a logarithmic removal model with an R value of 0.99 (Figure 28). The experimental constructed wetlands removed approximately 99

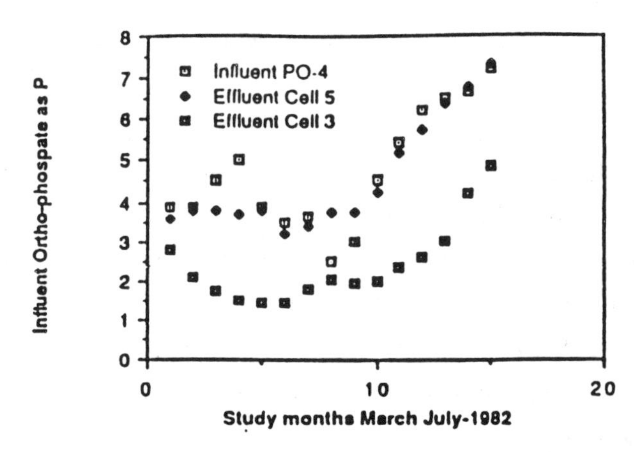

Figure 27. Phosphorus Removal from Cell 3 (0.5 gpd/ft²) and Cell 5 (2.94 gpd/ft²)

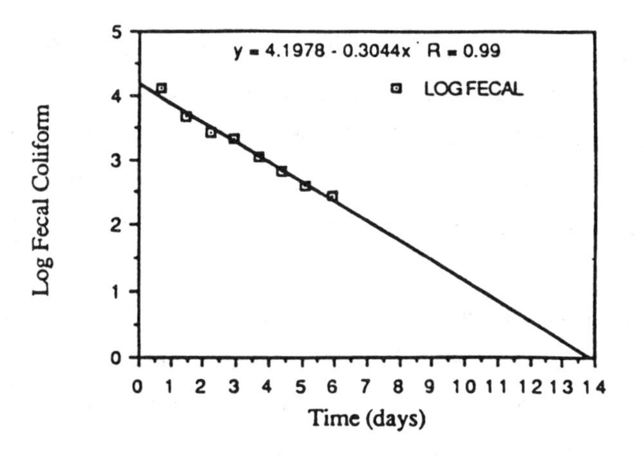

Figure 28. Fecal Coliform Removal in Pilot Project Cell 8, 1985-1986

percent of the fecal coliform in 6 days of retention. Extrapolated data showed a 3 log removal (99.9) after 10 days of retention.

Engineering Approach. Wherever possible, emphasis should be placed on a "low impact" engineering approach. This not only avoids unnecessary expense but can enhance the natural processes involved in the use of wetlands. The design concept should be to have the system fit naturally into the landscape following the topography and minimizing straight dikes and 90° corners. If wetland habitat is a value, then islands with nooks and crannies should be liberally placed in the wetland surface.

Water Depth. Initial establishment of emergent plants in surface-flow wetlands may require initial shallow water depths of 0.15 m (Gearheart et al., 1982), after which depth should be increased to 0.3-0.6 m. The water depth should be raised as the emergent macrophytes put on stem length and increase in numbers of new sprouts from the rhizomes. If *Scirpus* is the main desired species, initial water depths should be maintained as shallow as possible to prevent *Typha* domination. *Scirpus* does well in -0.05 m to 3 m depth, whereas *Typha* dominates at depths >0.15 m (Stephenson et al., 1982).

Cell Construction. Bottom contours should be smooth, and abrupt bathymetric discontinuities should be avoided to minimize potential problems with short circuiting, and to avoid formation of refuges for predators during periods of drawdown (Fog et al., 1982). The cells should be designed with length to width ratios of 5:1 to 10:1 and constructed in the direction of the major axis of flow. The height of embankments should be as small as practicable to maximize wind fetch, thus augmenting aeration and restricting problems with duckweed.

Drainage Points Within Cells. It will be important to provide for some means of enabling complete wetland drainage. This could involve pumps or internal drainage points. If drainage points are used, it will be important to ensure that they are readily able to be located if required in the future. It may be important to be able to take separate cells out of commission for maintenance, if necessary. Therefore, wherever practicable in final design, attention should be given to ensuring cells are self-contained, and that effluent bypasses will be able to be implemented.

Inlet Systems. Knight (1987) concluded that a point source discharge into a wetland would be preferable to a wide-front discharge. Nevertheless, on the basis of an overall review of available data, including discussions in the United Kingdom, it is concluded that the discharge into a wetland should be made across a wide front to take advantage of initial BOD and SS reductions in the first 10-15 m of a wetland. This should also avoid the rapid formation of a BOD or SS front which is referred to by Knight (1987).

Outlet Systems. The outlet system should be designed to encourage a uniform collection of the effluent across the outlet zone of the wetland. This usually would involve several outlet weirs connected by means of a collective manifold. The velocity is extremely low in the marsh, except in the outlet region. If too much flow is forced through one collection point, increased velocities will cause short circuiting (Gearheart et al., 1984).

Rhizome Planting. Rhizomes of most wetland plants are suitable for use in propagation. It is less resource intensive to spread rhizomes than to plant individual plants. However, rhizomes are themselves sensitive to desiccation and should be carefully managed. When obtaining wetland plant rhizomes, or individuals for planting, stocks should be obtained from the immediate vicinity of the wetland site, if possible.

Planting clumps of emergent plants (1 to 2 ft^2) in the late fall with their vegetative tops cut off on a 1 m^2 centering will give a constructed wetland a good start. These clumps will bring with them sufficient soil and enough rhizome protection to endure a running start in the spring.

The advocated approach would be "plug planting" with emergent wetland plants in early- to mid-autumn and allow approximately one year for suitable establishment of the plants. However, it is possible to plant at any time of the year provided that rhizomes and plants are kept constantly moist.

Soil Composition. The soils used for a wetland plant establishment should be well-tilled fertile humus/clayey soils. These types of soils will allow for rapid germination of plants and for optimum rhizome proliferation (Gearheart et al., 1986).

At Listowel, the wetland soil was conditioned prior to planting (Herskowitz et al., 1987). In that case, marsh basins were composed of compacted clay, filled to a depth of 0.15 m with a combination of topsoil and peat (10 percent by volume). It might be advisable, for example, to utilize dewatered sewage sludge as a soil conditioner. This would be disked into the soil prior to planting. This soil conditioning will be required if soil is to be removed from the wetland area for construction of embankments.

Temperature. Wetlands will operate at higher efficiencies in situations of higher temperatures but will operate over a wide temperature range. This is related to plant physiology and the amount of litter deposited by plants during seasonal winter dieback. This would account for the apparent success of wetlands in Saskatchewan, Canada, where temperatures were well below freezing during the winter months (Laksman, 1982).

Botanical Input. In designing wetlands, the engineering principles of cell construction, flow direction, depth, inlet and outlet structures, and application rates are relatively straightforward. However, the key to successfully establishing a wetland system lies in the installation and maintenance of the wetland plants. It takes at least two full growing seasons for the plant density to be great enough to have an effect. This assumes a relatively high initial planting density and high plant survival.

Plant Species Suitability. In terms of species suitability *Typha* (cattail) is considered to be less suitable than *Scirpus* (bulrush), for a number of reasons. *Typha* has greater capacity for oxygen translocation to the root system, greater degree of litterfall, which can cause problems with anaerobic conditions, and problems with windthrow if its roots are not adequately deep in the soil.

Odors. Hydrogen sulfide generation and associated odors occur periodically at wetland outlets. The incidence of this problem is increased by use of plants which lead to anaerobic conditions (e.g., *Typha*), and in these cases odor problems are exacerbated when the effluent is discharged in a cascade. Controlling the outlet can prevent hydrogen sulfide odor problems.

The odors identified most commonly in aquatic wetland treatment systems are associated with organic compounds containing sulphur, such as mercaptans and skatoles, and with hydrogen sulfide. Hydrogen sulfide is produced by obligate anaerobic organisms capable of reducing sulfate (Tchobanoglous et al., 1979). In the absence of oxygen and nitrate, sulfate will serve as an electron acceptor and is reduced to hydrogen sulfide in the process. Thus the presence of sulfate in the wastewater can lead to the formation of hydrogen sulfide in the bottom sludge accumulations. The organic matter in the sludge accumulation serves as a carbon source for the anaerobic process. The incomplete oxidation of other organic materials containing sulphur will also lead to the development of odors.

Anaerobic conditions develop when the treatment process is overloaded organically. Most commonly, anaerobic conditions develop near the effluent end of an aquatic treatment system.

Strategies that can be used to control the development of odors include the following: more effective pretreatment to reduce the total organic loading on the aquatic treatment system, more effective effluent distribution, step feed of influent waste stream, and supplemental aeration.

In constructing wetlands, it will be necessary from the outset to consider whether the organic loading will be likely to cause odor problems. The most common approach is to estimate BOD loading in terms of kg/ha/day and compare with published BOD loading rates (e.g., as outlined in Knight, 1987). Organic loadings at 166 kg/ha/day or less proved to be most effective in terms of not overloading the Arcata wetland system (Gearheart et al., 1983). Interestingly, loadings at Listowel (Herskowitz et al., 1987) were in the range 0.27-0.92 kg/ha/day and hydrogen sulfide generation was observed, two to three orders of magnitude less than Arcata.

Final Segment Polishing. The final segment of the wetland is of major significance in terms of final effluent polishing and retention of suspended solids. This final segment should be retained in as undisturbed a state as possible. If harvesting is selected as a management option (note that research has shown harvesting to be not necessary and even deleterious), then the final section of the wetland should be retained in this natural state at all times. Harvesting 10 to 20 percent of the wetland cell per year can be an effective way to manage vegetation without affecting water quality in the effluent. Care should be taken to minimize the disruption of the vegetation in the near vicinity of the effluent zone (Gearheart et al., 1985).

Summary. Wetland treatment effectiveness is a function of retention time and capacity of the vegetation and sediments to retain and/or cycle certain constituents (Gearheart et al., 1985). In using a wetland to polish domestic secondary treated effluent, the following general guidelines are considered reliable. It has been shown that an effluent suspended solids level of 5 to 10 mg/L can be achieved with a retention period of about 1 to 2 days. A longer retention time is required for effective BOD removal. An effluent BOD value of 10 to 15 mg/L can be achieved with 4 to 8 days of retention of a secondary treated effluent. Total nitrogen levels of the order of 4 to 6 mg/L can be achieved with 10 to 12 days of retention. Total phosphorus levels of 2 to 4 mg/L can be achieved with 15 to 20 days of retention. In the case of nitrogen and phosphorus removal vegetation and detritus, harvesting and collection will be necessary prior to decomposition to capture the nitrogen and phosphorus associated with the biomass. This management interval will be a variable depending on the removal requirements, the growing period, and the size of the wetland. In the long run, when steady state conditions are reached, an annual harvesting schedule of a portion of the wetland will be required.

REFERENCES

Brix, H. and H. Schierup H. 1987. "Double Benefits of Marcophyte Revival," Water Quality International, 2:22-23.

CH_2M Hill. 1985. "Wetland Treatment and Landscape Irrigation for Wastewater Reuse at the South County Regional Park, Palm Beach County, Florida." Report for Palm Beach County Utilities, West Palm Beach, Florida.

CH_2M Hill. 1986. "Boggy Gut Wetland Treated Effluent Disposal System, Hilton Head, South Carolina—Final Status Report November 1986." Report for Sea Pines Public Service District.

Cooper, P.G. and A.G. Boon. 1987. "Use of Phragmites for Wastewater Treatment by the Root Zone Method." Paper presented to NE Branch of IWPC, February 25.

Crites, R.W. and T.J. Mingee. 1987. "Economics of Aquatic Wastewater Treatment Systems," pp. 879-888. In: Aquatic Plants for Water Treatment and Resource Recovery, K.R. Reddy and W.H. Smith (eds.), Magnolia Publishing Inc.

DeJong, J., T. Kok, and A.H. Koridon. 1985. "The Purification of Wastewater and Effluents Using Marsh Vegetations and Soils," Proc. EWRS 5th Symp. on Aquatic Weeds.

Demgen, F.C. 1985. "An Overview of Four New Wastewater Wetlands Projects," pp. 579-595. In: Future of Water Reuse, Vol. 2, Proc. Water Reuse Symp. III, Aug. 26-31, San Diego, California, AWWA Research Foundation.

Feeney, P., B. Morrel, D. Click, and J.A. Jackson. 1986. "Wetlands Wastewater Disposal: A Simple, Environmentally Safe Solution," Report to Florida Section AWWA, Florida PCA, Florida Water and Pollution Control Operators Association. West Palm Beach, Florida, Nov. 11-14.

Fetter, C.W. Jr., W.E. Sloey, and F.L. Spangler. 1978. "Use of a Natural Marsh for Wastewater Polishing," pp. 290-307, Journal WPCF, February.

Fog, J., T. Lampio, J. Rooth, and M. Smart. 1982. "Managing Wetlands and Their Birds - A Manual of Wetland and Waterfowl Management," Proc. 3rd Technical Meeting on Western Palearctic Migratory Bird Management, held at the Biologische Station Rieselfelder Munster, FRG 12-15 (Oct.), Publ. Int. Waterfowl Research Bureau.

Gearheart, R.A., S. Wilbur, J. Williams, D. Hull, et al. 1982. City of Arcata Marsh Pilot Project Second Annual Progress Report, Sept. 1981. Report to California State Water Resources Control Board, August.

Gearheart, R.A., S. Wilbur, J. Williams, D. Hull, B. Finney et al.. 1983. Final Report City of Arcata Marsh Pilot Project, City of Arcata Department of Public Works, Arcata, California, (April).

Gearheart, R.A., J. Williams, H. Holbrook, and M. Ives. 1985. City of Arcata Marsh Pilot Project Wetland Bacteria Speciation and Harvesting Effects on Effluent Quality. Environmental Resources Engineering Department, Humboldt State University, Arcata, California.

Gersberg, R.M., B.V. Elkins, and C.R. Goldman. 1983. "Nitrogen Removal in Artificial Wetlands," Water Research, 17 (9):1009-1014.

Gersberg, R.M., S.R. Lyon, B.V. Elkins, and C.R. Goldman. 1984. "The Removal of Heavy Metals by Artificial Wetlands," pp. 639-648, In: Future of Water Reuse, Vol. 2, Proc. Water Reuse Symp. III, Aug. 26-31, San Diego, California, AWWA Research Foundation.

Gersberg, R.M., B.V. Elkins, and C.R. Goldman. 1984. "Use of Artificial Wetlands to Remove Nitrogen from Wastewater," Journal WPCF, 56 (2) February.

Gersberg, R.M., B.V. Elkins, and C.R. Goldman. 1984. "Wastewater Treatment by Artificial Wetlands," Water Science and Technology, 17:443-450.

Gersberg, R.M., R. Brenner, S.R. Lyon and B.V. Elkins. 1987. "Survival of Bacteria and Viruses in Municipal Wastewaters Applied to Artificial Wetlands," pp. 237-247. In: Aquatic Plants for Water Treatment and Resource Recovery, K.R. Reddy and W.H. Smith (eds.), Magnolia Publishing Inc.

Gersberg, R.M., V.G. Elkins, S.R. Lyon, and C.R. Goldman. 1986. "Role of Aquatic Plants in Wastewater Treatment by Artificial Wetlands," Water Research, 20 (3):363-368.

Greiner, R.W. and J. DeJong. 1982. "The Use of Marsh Plants for the Treatment of Wastewater in Areas Designated for Recreation and Tourism," Flevobericht No. 225, Introductory paper presented at 35th International Symposium (Cebedeau) May 24-26, at Liege.

Hammer, D. and R.H. Kadlec. 1983. Design Principles for Wetland Treatment Systems, EPA 600/S2-83-026, May.

Herskowitz, J., S. Black, and W. Lewandowski. 1987. Listowel Artificial Marsh Treatment Project, pp. 237-246. In: Aquatic Plants for Water Treatment and Resource Recovery, K.R. Reddy and W.H. Smith (eds.), Magnolia Publishing Inc.

Howard-Williams, C. 1985. Cycling and Retention of Nitrogen and Phosphorus in Wetlands: A Theoretical and Applied Perspective, Freshwater Biology 15:391-431.

Ives, M.A. 1986. "The Fate of Natural Virus in an Artificial Marsh Wastewater Treatment System Utilizing a Coliphage Model." Master's thesis. Humboldt State University, Arcata, California, 85 pp.

Kadlec, R.H. 1979. "Wetland Tertiary Treatment at Houghton Lake Michigan," pp. 101-139. In: Bastian, R.K. and S.C. Reed (eds.). Aquaculture Systems for Wastewater Treatment: Seminar Proceedings and Engineering Assessment, EPA 430/9-80-006.

Kappel, W.M. 1979. "The Drummond Project - Applying Lagoon Effluent to a Bog: An Experimental Trial," pp. 83-90. In: Bastian, R.K. and Reed, S.C. (eds.). Aquaculture Systems for Wastewater Treatment: Seminar Proceedings and Engineering Assessment, EPA 430/9-80-006.

Knight, R.L. 1985. "Wetlands: An Alternative for Effluent Disposal, Treatment, and Reuse," pp. 6-9, Florida Water Resources Journal, (Nov.-Dec.).

Knight, R.L. 1987. Effluent Distribution and Basin Design for Enhanced Pollutant Assimilation by Freshwater Wetlands, pp. 913-921. In: Aquatic Plants for Water Treatment and Resource Recovery, K.R. Reddy and W.H. Smith (eds.), Magnolia Publishing Inc.

Laksman, G. 1982. "Natural and Artificial Ecosystems for the Treatment of Wastewaters, Saskatchewan Research Council Publication No. E-820-7-E-82.

Lacy, J. 1983. "A Bathe Study of Copper, Lead, and Zinc in a Marsh System - City of Arcata, California," Special Study Environmental Resources Engineering Department, Humboldt State University, Arcata, California.

Lienard, A. 1986. Study of the Sewage Purification Works Using Beds of Macrophytes at Logis a St. Bohaire, French Ministry of Agriculture, Department of Water Quality, Fishing and Fish Farming, Lyon, October.

Nelson, R.W. and E.C. Weller. 1985. A Better Rationale for Wetland Management, Environmental Management, 8:295-308.

Nichols, D.S. 1983. Capacity of Natural Wetlands to Remove Nutrients from Wastewater, Journal WPCF, 55 (5) (May).

Phillips, G.L., B. Ayling, C. Clarke, and C. Thomas. 1987.

Rossiter, J.A. and R.D. Crawford. 1984. Evaluation of Artificial Wetlands in North Dakota: Recommendations for Future Design and Construction, Transportation Research Record 948.

Stephenson, M., G. Turner, P. Pope, J. Colt, A. Knight, and G. Tchobanoglous. 1982. Publication No. 65. The Use and Potential of Aquatic Species for Wastewater Treatment Appendix A: Environmental Requirements of Aquatic Plants, California State Water Resources Control Board, Sacramento, California.

Stowell, R., R. Ludwig, J. Colt, and G. Tchobanoglous. 1981. Concepts in Aquatic Treatment System Design, pp. 16555-16569, Journal of the Environmental Engineering Division, Proceedings of the American Society of Civil Engineers, 107(EE5) October.

Tchobanoglous, G. 1987. "Aquatic Plant Systems for Wastewater Treatment: Engineering Considerations, pp. 27-48. In: Aquatic Plants for Water Treatment and Resource Recovery, K.R. Reddy and W.H. Smith (eds.), Magnolia Publishing Inc.

Tchobanoglous, G., R. Stowell, R. Ludwig, J. Colt, and A. Knight. 1979. Aquaculture Systems for Wastewater Treatment: Seminar Proceedings and Engineering Assessment, EPA 430/9-80-006, pp. 35-55. In: Bastian, R.K. and Reed, S.C. (eds.).

Valiela, I., S. Vince, and J.M. Teal. 1976. "Assimilation of Sewage by Wetlands." In: Estuarine Processes, Vol. I, M. Wiley (ed.), Academic Press, 234-253.

Wile, I., G. Palmateer, and G. Miller. 1981. Use of Artificial Wetlands for Wastewater Treatment, pp. 255-271. In: Proceedings of the Midwest Conference on Wetland Values and Management, B. Richardson (ed.), St. Paul, Minnesota, June.

Williams, T.C. and J.C. Sutherland. 1979. Engineering, Energy and Effectiveness. Features of Michigan Wetland Tertiary Wastewater Treatment Systems, pp. 141-173. In: R.K. Bastian and S.C. Reed (eds.) Aquaculture Systems for Wastewater Treatment: Seminar Proceedings and Engineering Assessment, EPA 430/9-80-006.

Constructed Wetlands Experience in the Southeast

Robert J. Freeman Jr., PE

Cobb County Water System
Marietta, Georgia

Constructed wetlands (CW) used as wastewater treatment technology have seen a dramatic increase in use throughout the United States in the last several years. In part due to the warmer climate and availability of land for siting, CW systems have been utilized in the "sun-belt" states more frequently than elsewhere. Of the 154 constructed wetlands systems (operational, under construction, and planned) inventoried in the United States in March 1990, by Sherwood Reed under contract to EPA, 97 were in EPA Regions 4 and 6, the southeast and southwest. In spite of the number of these systems in use, the shortage of meaningful data and the lack of understanding regarding the basic physical and biochemical processes taking place has resulted in little progress towards a sound design approach. This problem has led to unexpected difficulties with a number of the CW systems in operation. Both gravel substrate, subsurface flow, and soil substrate, surface flow, systems have encountered serious problems, in some cases jeopardizing the continued use of the CW technology at those locations. This paper will discuss some of those issues and the measures being considered to attempt to avoid those problems in the future.

Constructed wetlands have been generally grouped into two basic categories, the simplest being the systems in which rooted aquatic plants are planted in a soil substrate within a constructed earthen basin. The basin may be lined or not depending on natural soil permeability and ground-water protection requirements. The systems are generally designed to allow wastewater effluent following preliminary treatment to flow at a depth of 1 to 2 in. up to 12 to 18 in. through the basin in a plug flow pattern. The Water Pollution Control Federation (WPCF) design manual titled "Natural Systems for Wastewater Treatment" designates these systems as Free Water Surface (FWS) systems referring to the nature of the surface flow. The second type of CW system is similar to the FWS systems except the basin is filled with aggregate such as gravel or crushed stone to a depth of 12 to 24 in. in which the aquatic vegetation is planted and through which the wastewater flows with no visible surface flow. These systems have likewise been designated as Vegetated Submerged Bed (VSB) systems by the WPCF design manual. A brief comparison of those of the two types of systems is shown in Table 9.

Table 9. Comparison Of FWS And VSB Systems

FWS Systems	**VSB Systems**
Lower installed cost/gal	Greater assimilation rate — less land required
Simpler hydraulics	No visible flow — less nuisance, vector problems, odors
More natural wetland values can be incorporated into the system	More cold tolerant

The division of the 97 CW systems in Regions 4 and 6 is 17 of the FWS type and 80 of the VSB type. The prevalence of the VSB type system is due, at least in part, to the success Dr. Bill Wolverton, formerly with NASA at their southwest Mississippi test facility, had in researching and promoting the use of these systems. Of those 80 VSB systems identified, 49 are located in Mississippi and Louisiana.

The original and on-going popularity of the CW technology derives primarily from its two-fold promise of lower costs and little requirement for operation and maintenance compared to conventional technology. The EPA construction grants program encouraged the use of the technology due to the Innovative/Alternative (I/A) grant bonus received by grantees selecting the CW systems of either type. The Tennessee Valley Authority (TVA) has also played a significant role in encouraging these systems, especially in the Tennessee Valley area.

Benton, Kentucky, was assisted by TVA in construction of a CW system to treat the wastewater from the approximately 4,600 residents. Their previous system was a 10.5 ha (26 acre) facultative two-cell lagoon with a flow as high as 4,500 m^3/d (1.2 mgd) due to infiltration/inflow. The original 4.0 ha (10 acre) second cell was converted to three equal sized parallel cells, one filled to a depth of 0.6 m (2 ft) with crushed limestone and the other two left as native soil, giving one VSB cell and two FWS cells (Figure 29). The VSB cell was planted with softstem bulrush and the FWS cells were planted with woolgrass in one and arrowhead and cattail in the other. The system was designed for an average flow of 4,160 m^3/d (1.1 mgd) with 50 percent of the flow intended to go through the VSB and the remaining 50 percent to be equally divided between the two FWS cells.

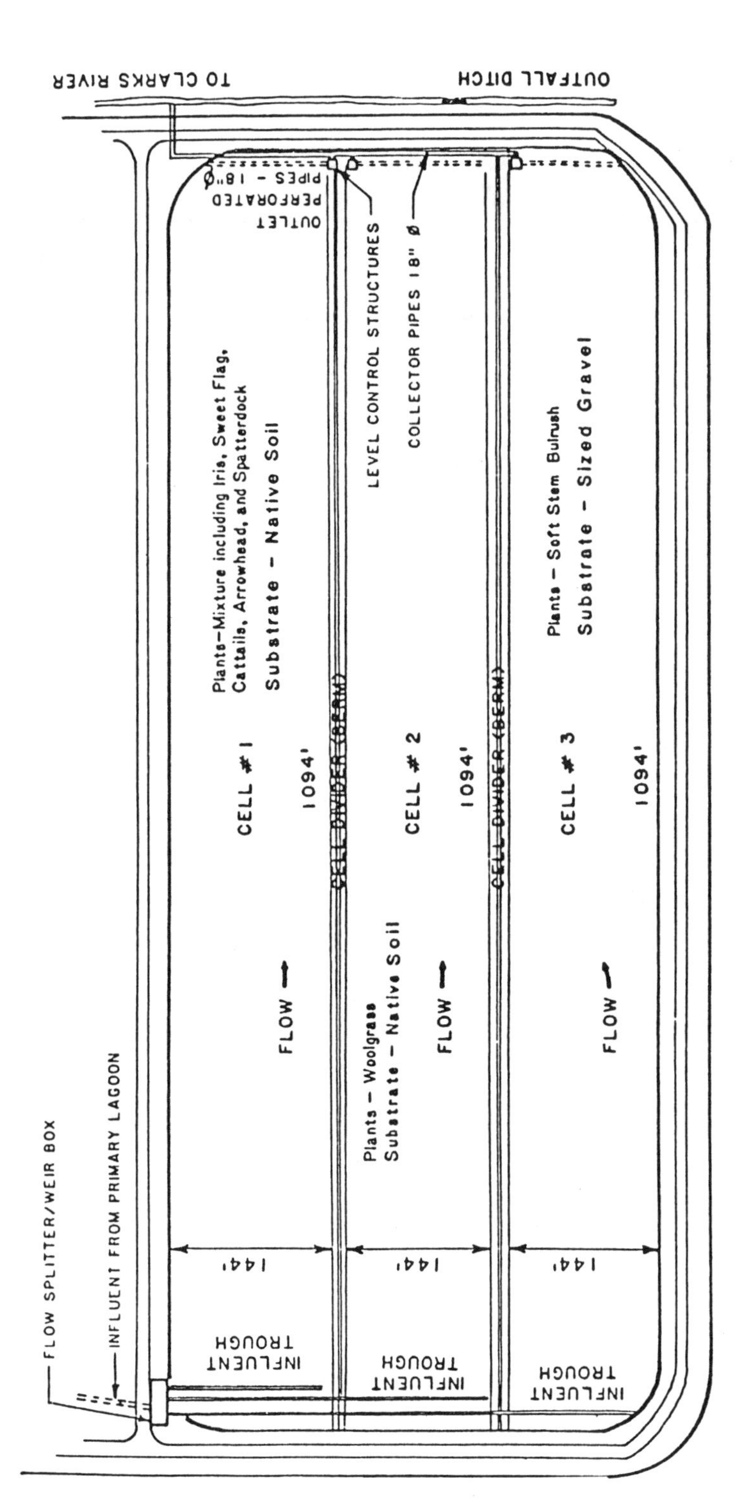

PLAN VIEW

BENTON, KY.

CONSTRUCTED WETLANDS SEWAGE TREATMENT SYSTEM

GRAVEL AND SURFACE FLOW MARSHES

Figure 29. CW Sewage Treatment System—Gravel and Surface Flow Marshes

The City's NPDES permit required the following limits to be met:

	BOD (mg/l)	TSS (mg/l)	NH_3 (mg/l)	DO (mg/l)
SUMMER	25	30	4	7
WINTER	25	30	10	7

During the construction of the wetland system all of the lagoon effluent was routed through the VSB cell while the FWS cells were built. During the first year of full operation, 1988, the influent flows to the wetland cells averaged a BOD_5 of 25 to 30 mg/L, TSS of 40 to 70 mg/L, NH_3 of 6 to 9 mg/L and flows averaging 1,800 to 2,600 m^3/d (0.5 to 0.7 mgd). The flows were within the design expectations with the exception of rainfall related high flows and the construction period diversion of all the flow through the VSB as mentioned.

The BOD_5 removals experienced were quite good, averaging 50 percent to 60 percent in the FWS cells and 80 percent in the VSB cell. The NH_3 performance, however, was very disappointing. All three cells experienced a significant increase in NH_3 through the cells, on the order of 30 percent to 50 percent, yielding effluent consistently in noncompliance with the summer NH_3 limits and occasionally exceeding the winter NH_3 limit as well. The effluent dissolved oxygen (DO) limits were also consistently violated, both summer and winter. This pattern continued in 1989, and in 1990 two major changes were made to the Benton system to attempt to remedy the lack of nitrification occurring. The first attempted solution was the implementation of a recycle system in one of the FWS cells to attempt to increase the availability of DO to provide an environment in which the nitrification reactions could occur. This effort proved fruitless even at a 3 to 4Q recycle and using sprinklers to introduce DO. The second change was to reduce the loading in the VSB and one of the FWS cells dramatically and route the remaining flow through the remaining FWS cell. The original design of the FWS cells called for a loading of 1.5 ha/1,000 m^3/2d (14 ac/mgd). The reduced loading was set at 4.5 ha/1,000 m^3/d (42 ac/mgd). This reduced loading corresponds to the area required for ammonia conversion in an FWS cell as calculated using the formulations in the WPCF Natural Systems Manual. Unfortunately, the nitrification in that cell still did not appear to increase since the NH_3 in the effluent still exceeded that in the influent by 25 percent for the two months of data available.

It is not clear why the nitrification reactions are not more effective in reducing ammonia (some is being oxidized as the decrease in TN and TKN suggests that organic N is being converted to NH_3 at the same time that some nitrification is occurring). Possible deposition of organic material from the lagoon during the higher loading period may have created a "background" DO load that will take some time (months or years) to satisfy such that enough DO remains for more complete nitrification to take place. The often-cited ability of plant roots to "pump" oxygen down into the VSB bed also appears to be questionable. Several observation pits dug in the VSB cell showed little root penetration below the 0.15 to 0.25 m (6 to 10 in.) depth, contrary to the expected full depth root penetration. It also may be that short-circuiting could be occurring in the FWS cell that effectively reduces the area and increases the net loading such that nitrification is not encouraged. Remedying any of these possible problems would at best be difficult.

The VSB cell also experienced surfacing of wastewater at the original design flow of 2,080 m^3/d (0.55 mgd). A detailed investigation of the cause of the surface flow showed that the crushed limestone bed had plugged significantly with an inorganic gel-like substance formed from a reaction of the limestone and silicon and algae in the wastewater. A reduction in loading to approximately 10 percent of design brought the water level back below the surface. The resulting loading was 11.7 ha/1,000 m^3/d (105 ac/mgd) compared to the original design of 0.7 ha/1,000 m^3/d (6.5 ac/mgd). The VSB cell is now finally achieving satisfactory nitrification, with effluent NH3 values of 3 mg/L obtained (within the permit limit of 4 mg/L). At this loading rate, however, another 90 acres or so of VSB cells would be necessary to bring Benton into reliable compliance with their ammonia limit; this would be clearly economically prohibitive.

These results at Benton seem to be in line with the recommendations in the WPCF Manual of 1.2 ha/1,000 m^3/d (11.2 ac/mgd) of VSB cell and 3 ha/1,000 m^3/d (28 ac/mgd) of FWS cell for removal of BOD. For significant ammonia conversion the size of those cells would increase to at least 4 ha/1,000 m^3/d (40 ac/mgd). These results bring into serious question the commonly used assumptions regarding oxygen transport into the root zone in VSB systems, the loading rates used for those systems, and the effectiveness of CW systems for NH_3 removal except at very low loadings.

Denham Springs, Louisiana, is an example of a VSB system designed and constructed under commonly used design criteria that this author believes are in serious question. The system was designed to treat 11,300 m^3/d (3.0 mgd) average flow following a 32 ha (80 ac) two-cell facultative lagoon to meet effluent limits of 10 mg/L BOD and 15 mg/L TSS monthly. The CW system consists of 3 VSB cells of 2 ha (5 ac) each, 320 m (1,050 ft) long and 66 m (215 ft) wide. The aggregate consists of 0.45 m (18 in.) of 0.5 to 1.8 cm (1 to 4 in.) limestone topped with 0.15 m (6 in.) of 2.5 cm (1 in.) gravel. Bulrush and canna lilies are planted in the cells. In 1989, the flow to the system averaged 9,000 m^3/d (2.4 mgd) with an influent BOD of 27 mg/L. During that period the 10 mg/L BOD permit limit was exceeded about half the time and a significant portion of the length of the cells have experienced surfacing of wastewater. The design loading of this system is approximately 0.5 ha/1,000 m^3/d (5.0 ac/mgd) at average flow, about 20 percent heavier than Benton, Kentucky, and about twice as heavy as the WPCF manual recommendation. It will be interesting to see how the performance of the Denham Springs system fares as it reaches design flow.

Two other troubling concerns with VSB systems are the use of the larger size rock in the apparent hope that plugging will not be a problem, and the large length/width ratios. Using the design methodology in the WPCF manual, as the size of the aggregate goes up and the corresponding porosity goes down, larger area cells are required to give adequate opportunity for the biological processes to occur. In addition, the configuration of VSB cells should generally be much wider and not as long, a result based loosely on Darcy's Law. A system like Denham Springs, if designed along the lines of the WPCF Manual for BOD removal only, would have the configuration shown in Table 10.

Table 10. Denham Springs System
Denham Springs, LA: Q average = 11,355 m^3/d (3 mgd)

	Existing Design	New Design
Loading rate	0.5 ha/1,000 m^3/d (5 ac/mgd)	1.6 ha/1,000 m^3/d (15 ac/mgd)
Area needed	6 ha (15 ac)	18 ha (45 ac)
Length x width	320 m x 197 m (1,050 ft x 645 ft)	80 m x 2,270 m (262 ft x 7,440 ft)

It is obvious that a dramatically different design would result if the WPCF procedure were used. The two-dollar question is—which one is correct? Since the WPCF approach would require a 300 percent increase in land and in rock (the most expensive thing about the system), its use would render the economics of the VSB as described above questionable as a viable alternative. The use of smaller aggregates would reduce the land area significantly and might bring the costs back to an acceptable level but would still result in much wider and shorter cells. If the design were to provide for NH_3 removal, however, the size required would increase significantly. An FWS system, for comparison, would require approximately 25 ha (62 ac) for 11,355 m^3/d (3 mgd) using the WPCF approach for BOD removal but would increase to 70 ha (172 ac) for NH_3 removal.

The sobering implication of these large differences in the design of the Denham Springs system and the WPCF approach is that Denham Springs is typical of the design approach at a number of other operational or under construction systems in Louisiana as well as several in nearby states. An EPA-funded research effort presently underway is intended to evaluate more thoroughly some of these VSB systems and shed some new light on these systems.

A promising VSB system designed and constructed by TVA has been in operation since late 1988 at Phillips High School in Bear Creek, Alabama. The system is designed to treat 76 m^3/d (20,000 gpd) of package treatment plant effluent to limits of BOD=20 mg/L, TSS=30 mg/L, and NH_3 = 8 mg/L. The VSB is sized at 0.2 ha (0.5 ac) resulting in a loading of 2.6 ha/1,000 m^3/d (25 ac/mgd) with 0.3 m (12 in.) of pea gravel substrate. This loading rate is one-fourth the original design of the Benton, Kentucky, VSB cell and less than one-fifth the loading of the Denham Springs VSB, in spite of the fact that the influent wastewater is more highly treated (influent BOD averaged 13 mg/L, NH_3 10.7 mg/L). The performance of the Phillips VSB is excellent, as shown in Table 11.

Table 11. Phillips High School VSB

	BOD mg/L	nTSS mg/L	NH_3 mg/L
INFLUENT	13	60	10.7
EFFLUENT	<1	<3	1.8

The high degree of nitrification is due, at least in part, to the low loading of the system. The nature of operation of the Phillips VSB may also contribute to its high-level treatment since the high school is not in session for most of the summer and, therefore, little or no loading is contributed then. Low loading combined with corresponding low nutrient application and a relatively shallow gravel depth could be responsible for the full bed root penetration that has been observed there, unlike that observed at several other systems. This intermittent type of operation may assist the system in staying aerobic so the nitrification reactions can take place, as in an intermittent sand filter or the like.

In FWS systems, two beneficial modifications have been utilized in the design and construction of the cells that may address the problems of short-circuiting and inadequate oxygen transfer. The Fort Deposit, Alabama, FWS system has two parallel cells that are subdivided by three deep zones that are approximately 1 m (3 ft) deeper than the rest of the cell to prohibit plant growth and redistribute the flow across the cell, shown in Figure 30. The zones are 5 to 8 m (15 to 24 ft) wide and may also serve as a recreation area to help keep the DO elevated. Fort Deposit is designed to treat 900 m^3/d (0.24 mgd) following a partially aerated lagoon to meet monthly limits of BOD=10 mg/L, NH_3=2 mg/L on a year round basis. The system has a total FWS area of 6.1 ha (15 ac) for a loading of 6.8 ha/1,000 m^3/d (60 ac/mgd). This rate is in line with the WPCF approach, due to the fact that Dr. Robert Knight of CH_2M-Hill designed the Fort Deposit FWS system and is the author of the Wetlands Systems chapter of the WPCF manual. The Fort Deposit system will begin a detailed performance evaluation this summer (1991), and results of the modifications utilized there will be forthcoming. Another FWS system designed by Dr. Knight for West Jackson County, Mississippi, incorporates the deep zones and also uses a shallow inlet area to maximize oxygen transfer. The FWS system is a 2,300 m^3/d (0.6 mgd) two cell (series) basin using 8.9 ha (22 ac) for a loading of 3.9 ha/1,000 m^3/d (27 ac/mgd). The inlet zone in each cell is very shallow, 5 cm (2 in.), gradually deepening to a normal depth of 0.3 to 0.5 m (12 to 18 in.). This inlet area should increase the oxygen transfer capability to enhance nitrification to meet the effluent limits of BOD=10 mg/L, NH_3=2 mg/L monthly on a year round basis. Since the loading rate in this system is relatively high for a nitrification system with limits this tight, the data collection effort, which will begin in earnest this summer (1991), will be interesting.

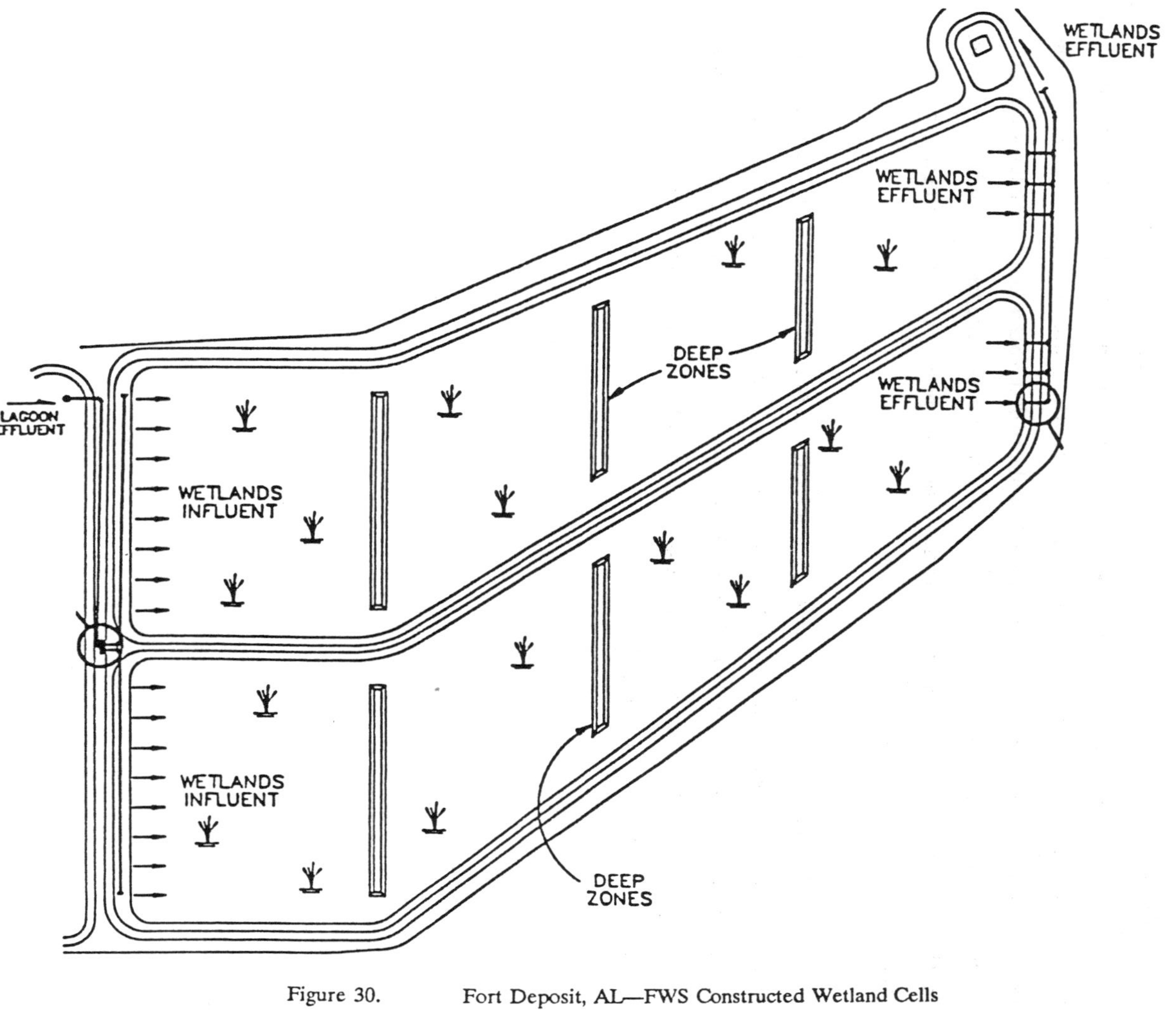

Figure 30. Fort Deposit, AL—FWS Constructed Wetland Cells

The story of CW systems in the southeastern part of the United States is obviously not over yet. It is also too soon to tell whether or not it will have a happy ending for the cities and towns that have built and are operating these systems. The evidence so far suggests that the early optimism which produced the relatively heavily loaded VSB and FWS systems was not warranted and a serious reevaluation of those systems may be required. Hopefully, the newer developments in CW system design and construction will prove to be successful in remedying some of the problems described.

Until more definitive information is available regarding design protocols and expected performance, the people involved in the review, approval, design, and construction of these systems must exercise caution in their use. While under the right circumstances a CW system can be a very desirable treatment technology, care must be taken to ensure an appropriate system for the situation is selected.

MUNICIPAL WATER USE EFFICIENCY

How Efficient Water Use Can Help Communities Meet Environmental Objectives

Stephen Hogye

U.S. Environmental Protection Agency
Washington, DC

Problem. This research project addresses how reducing water demand through more efficient water use can help communities deal with a number of environmental problems, ranging from ground-water contamination to compliance with expensive drinking water treatment requirements.

Background. Of the many important functions performed by local units of government, one of the most fundamental is the development of water supplies and the collection and treatment of wastewater. The economic vitality of any community is heavily dependent upon the availability of water in acceptable quantity and quality to sustain a multitude of uses, ranging from industrial manufacturing to drinking and dishwashing. Meeting this objective is becoming increasingly difficult. Available water sources are commonly contaminated, and development of new sources carries a growing penalty in terms of high financial and environmental cost.

As water resources are subjected to higher levels of stress, local governments are finding efficient water use to be an attractive means of meeting legitimate needs without sacrificing lifestyles or compromising community development objectives. Improvements in technology have resulted in products such as efficient showerheads, toilets, and sprinkler systems that satisfy consumer needs while using considerably less water. In addition to conservation measures, water efficiency can be increased through recycling and wastewater reclamation and reuse.

Many communities have gained considerable experience in adopting water conservation as a response to drought or as a means of deferring expansion of drinking water and wastewater treatment facilities. However, the purpose of this research project is to explore the effectiveness of efficient water use toward achieving a range of additional environmental objectives.

Research Approach. This project employs analytical computer models as well as field data to evaluate the benefits of efficient water use under several common community scenarios. These four hypothetical communities help illustrate the usefulness of reduced water use for communities ranging in size, type of water supply and wastewater system, rate of growth, and the kinds of environmental problems they face. The scenarios also include the need to upgrade drinking water and wastewater treatment levels.

For each community scenario, several water efficiency options are considered, and average and peak volume reductions are estimated. This illustrates the range of use reduction responses available, and ways to distinguish what techniques may be most promising in a particular situation. The impacts of these efficiency options on water utility operations and costs are assessed, including technical, economic, rate, and institutional considerations. From this, more general conclusions can be drawn about the usefulness of particular water efficiency programs for a variety of environmental compliance problems.

Water Efficiency Techniques and System Responses. A wide range of water efficiency techniques is available to today's system manager. These techniques produce different results in terms of peak and average volumes, with correspondingly different effects on system operation and budget. Actual results also vary in response to many other factors specific to an individual community. Of greatest interest in this context is how to match the environmental issue with the program that will yield the most benefit.

Utility actions such as leak repair, metering, and pressure reduction, primarily affect base load volumes. Increasing plumbing efficiency for new and existing buildings, through regulation, installation or retrofit, also affects base load. Another whole group of conservation efforts aims to reduce peak or seasonal volumes by affecting outdoor use. Utility rate structures can be designed for a variety of effects. Commercial and industrial conservation efforts vary in result, depending on the industry targeted.

Reduced water demand and wastewater flows are not the entire story. Local utility managers, municipalities, and federal and state regulators are also concerned about how these reduced volumes may affect system operations and budget. The biggest institutional concern may be how an individual community assesses the impact of a water efficiency program, and how they implement it.

A utility's or community's economic concerns may include the impact of reduced water demand on utility rates and revenues, as well as the price of water for customers (especially larger users). The relative costs and benefits of water efficiency strategies may appear as system capital savings, deferred costs, operation and maintenance expenses, or savings from downsizing. Of particular concern for some smaller communities is how to account for possible lost revenue from successful water demand reduction efforts, further impeding ability to finance environmental improvements.

Study Findings. Preliminary results of this study suggest the following:

- Actual results of a water efficiency program are very specific to the community's particular water and wastewater needs, the type of efficiency measures employed, and a host of other factors. While it may be possible to identify the general situations where a community should consider water use efficiency, it is essential that each community develop its own community-specific analysis of the potential costs and benefits of alternative programs for its own particular system.

- Reducing peak water demand can reduce the potential for an existing aquifer contaminant plume to enter drinking water wells. A community might thus target peak use through an aggressive outdoor water use reduction program.

- The success of conservation in downsizing drinking water treatment equipment depends on the particular treatment and technology employed. For example, some drinking water treatment facilities for very small service populations cannot be scaled back further.

- For small drinking water systems facing rising treatment costs, conservation may still be useful. This may be especially true for communities with a flat fee system, a high summer peak, a relatively high growth rate, or potential groundwater contamination. Communities with very small service populations and rates based on actual use may need to carefully consider their rate structure and expected conservation savings to avoid potential revenue shortfalls from reduced volumes.

- Even without expansion or new treatment costs, utility conservation measures involving system repair and maintenance may pay for themselves in communities with older infrastructures. This may be more likely for water supply than for wastewater distribution systems, due to the relative contributions of stormwater and infiltration/inflow problems for some systems.

- Wastewater collection systems may experience problems such as odor, septicity, and clogging due to flow reduction. However, many wastewater facilities provide better treatment under reduced hydraulic loadings, and also experience some savings in overall operating costs.

EPA's Offices of Policy Analysis, Wastewater Enforcement and Compliance, and Ground Water and Drinking Water are supporting this project. The author also acknowledges the substantial contribution of the project workgroup and participating consultants.

Impact of Indoor Water Conservation on Wastewater Characteristics and Treatment Process—Phase I Study

Robert A. Gearheart

Environmental Resources Engineering Department
Humboldt State University
Arcata, California

Introduction. The objective of this phase of the study is to determine the change in influent wastewater characteristics (BOD and suspended solids) as a result of water conservation strategies in California. This is the first phase of a study which examines the impact of indoor water conservation on wastewater collections and treatment systems. This study developed from an analysis of the San Diego Point Loma's advance primary clarification process as it related to change in removal efficiency as a result of indoor water conservation.

Not all of the facilities were able to supply all the necessary data for the purposes of developing a profile of the treatment system. The following treatment plants were examined in this phase of the study.

- San Diego's Point Loma Plant - Southern California
- Santa Barbara's El Estero Plant - Central California
- Los Angeles - L.A. County Sanitation District Joint Water Pollution Control Plant - Southern California
- Goleta's Wastewater Reclamation Plant - Central California
- Contra Costa Sanitary District - Northern California
- Arcata - Northern California

These plants were selected based upon their size, location in the state, and type of treatment process. The Point Loma treatment plant is an advanced primary treatment system with anaerobic digestion of the solids. The Los Angeles plant is a primary plant with secondary treatment (conventional activated sludge) of a portion of the primary effluent. The solids from both these facilities are anaerobically digested. The Contra Costa treatment plant is an advanced secondary treatment plant.

Water conservation programs are being proposed and implemented on several levels and in several regions in the United States. The need to conserve water and/or to serve a greater population with a finite water resource has become a major issue in the water utility sector. Nowhere has this demand been greater than in California. The last five years' successive droughts have forced water wholesalers and retailers to reevaluate their policies and programs. While the arguments for conserving water appear to be apparent to the public, there is resistance to structural changes in plumbing, etc., in the utility business and in the engineering profession.

The public for the most part recognizes the need to reduce water uses, especially if 1) there is a savings in their water bill and 2) there is a savings in their energy bill. There are some other possible savings in the wastewater collection and treatment systems. There have been limited numbers of studies to identify these beneficial and adverse impacts which might occur and any economic consequence, either capital cost and/or operation and maintenance cost, that might be incurred. Examples of the potential impact of indoor water conservation on wastewater treatment and collection include:

Treatment Plant Efficiency:

- Increased mass suspended solids removal in the primary treatment process
- Increased mass BOD removal in the primary treatment process
- Increased gas production in anaerobic digesters
- Decrease in mixing energy requirements in activated sludge process
- Reduction of hydraulic transient condition due to inflow/infiltration and hourly/daily fluctuations in the influent flow

Collection:

- Increased hydraulic efficiency, especially during periods of high inflow/infiltration
- Increased solubilization of particulate BOD
- Increased solids separation

- Increased anaerobic breakdown of organics
- Increased hydrogen sulfide production, causing odors and corrosion

Water Conservation Techniques. This paper is interested in the effects of water conservation on wastewater treatment, and the most significant factor that can change the characteristics of a wastewater flow is household water conservation.

Water Use. The average house has a consistent pattern of water use for a typical day. For each member of the house, one shower, four flushes of the toilet, and any dish and laundry water together account for almost all the water that travels through indoor plumbing to the sewer line. Indoor domestic water use in the United States ranges from 40 to 150 gallons/capita/day, with an average of 70 gal/capita/day (1).

Wastewater. Typical wastewater flows from residential areas in the United States range from 30 to 130 gal/capita/day with an average of 65 gal/capita/day (1). However, other countries have different ranges for their typical flows. Typical concentrations for domestic raw sewage's BOD and suspended solids range from 150 to 200 mg/L (2).

Indoor Plumbing Devices

Toilets. The principal water saving device is the ultra low flush toilet at 1.6 gal/flush. The state of California enacted a law effective January 1, 1992, that changes the California plumbing requirement on all new (residential and commercial) construction to 1.6 gal/flush tanks. The state's law since 1983 required all new buildings to use up to a maximum of 3.5 gal/flush. Table 12 gives an accurate idea of how much water savings can be expected by improving the type of toilet installed.

Showers. A typical, nonconserving shower head can deliver as much as 8 gallons per minute, far more than is necessary. In comparison, a low-flow shower head, in order to meet standards, cannot exceed 3 gallons per minute.

Water Conservation Effects on Wastewater Treatment/Collection. There have been a number of studies done on the effects of conservation on sewerage processes. After the drought in

Table 12. Toilet Fixtures

Type of Toilet Savings	Gallons/Flush	Gallons/Toilet/Day	%
Nonconserving	5.5	33.0	0
Low-Flush	3.5	21.0	36
Ultra-Low-Flush	1.5	9.0	73

California during the mid-'70s, the EPA sponsored a study titled "The Effects of Water Conservation Induced Wastewater Flow Reduction - A Perspective." (4) A summary of the conclusions follows:

- Half of the wastewater treatment systems (17) surveyed had operational problems during drought conditions. However, none of the encountered problems significantly affected the operations of any of the systems.
- All of the systems surveyed operated continuously throughout the survey period, despite any problems encountered.
- There was no correlation between low-flow induced wastewater quality changes and the treatment plants' occasional BOD/TSS violations.
- The BOD and SS concentrations of the wastewater entering the treatment plant generally increased, while the concentration leaving the plant generally decreased during years of flow reduction. The efficiency of treatment plant removal of BOD and SS generally increased slightly.
- The two most common treatment plant factors influenced by wastewater flow reductions were energy and chemical uses.
- The operation and maintenance costs for the collection systems generally decreased slightly. At 50 percent flow reduction, there was a 3 percent cost reduction, probably due to energy savings from lift pumps.
- Chemical use differences varied from plant to plant, while energy use generally decreased. In some cases, an increase in chemical costs caused an increase in O&M costs, while in others, the energy savings outweighed the encountered increases.

A literature review on water conservation revealed that this activity will result in significant increases in an activated sludge process wastewater treatment plant's influent substrate concentrations (5). The authors concluded that "changes in substrate removal efficiencies resulting from increases in influent substrate concentrations (water conservation) range from zero to a few percent." They found that effluent BOD and COD concentrations change nearly proportionally with their corresponding influent values. It is noted that this occurrence might make it difficult for a plant to meet concentration-based discharge standards, but there is no evidence that mass loading discharges will be affected.

<u>Wastewater Characteristic Changes</u>. The city of Goleta, located in southern California adjacent to Santa Barbara, reflected the greatest impact of indoor water conservation on sewage flows. An analysis was made of their influent characteristic monthly average flow, BOD, and suspended solids for the period January 1988 through July 1990. The flows show that until about January 1989 the average monthly daily flow was about 6.5 mgd. At that time severe water conservation practices were initiated in the city. The reduction in indoor water use is shown as the sewage flows drop about 1.5 mgd the first year with a continuing reduction through the period of the data set resulting in about 3.0 mgd or 58 percent reduction in sewage flows in a 20-month period (Figure 31). This reduction was achieved with an aggressive implementation of low-flush toilets and to some unknown extent with the use of graywater for horticulture purposes. The BOD and TSS concentration in the influent wastewater averaged 230 and 200 mg/L, respectively, during this nonconservation period. Santa Barbara wastewater flows also showed a drop from 15 mgd to 12.5 mgd during the same period (Figure 32).

Table 13 shows the range of flows for the POTWs shown in this study. The variation in the influent TSS and BOD can be seen in Table 14 and Table 15. This again reflects the dilution effect of inflow/infiltration, exfiltration, and, in the case of Goleta, water conservation. The average TSS values ranged from 140 mg/L for Contra Costa county to 460 mg/L for Los Angeles. BOD influent concentrations for these cities ranged from 100mg/L at Contra Costa to 360 mg/L in Los Angeles (Table 16). The ratio of the influent SS/DOD ranged from 0.84 to 2.00.

<u>Model: Water Conservation Strategies/Wastewater Characteristics</u>. In order to gain insight into the possible effects of household water conservation on the characteristics of household wastewater, a model was developed. The model based its analysis on user-supplied, volume-per-use data for different contributing devices of various levels of conservation technology. For example, volume-per-use data for contributors such as toilets, showers, and garbage disposal is supplied for four levels of technology.

<u>Application</u>. This model was used to analyze one community as it underwent a conservation project at three rates of implementation. The initial community size was chosen to be 100,000 people, and grew at an annual rate of 2 percent. A slow, medium, and fast implementation rate was applied to the 20-year project. The difference between the rates lies in the level of installed

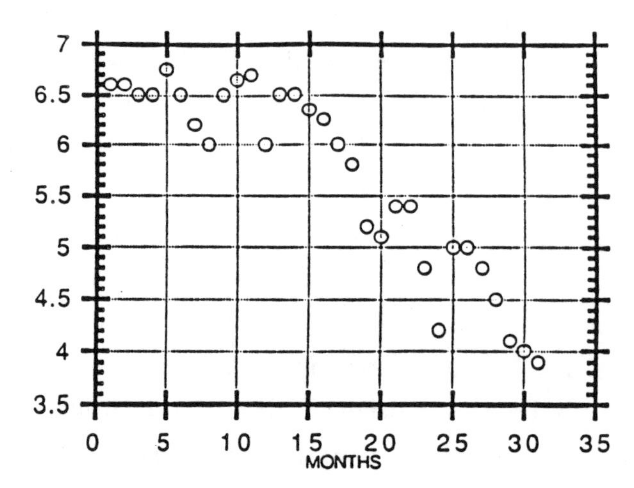

Figure 31. Influent Flow (mgd) Over Time for Goleta POTW

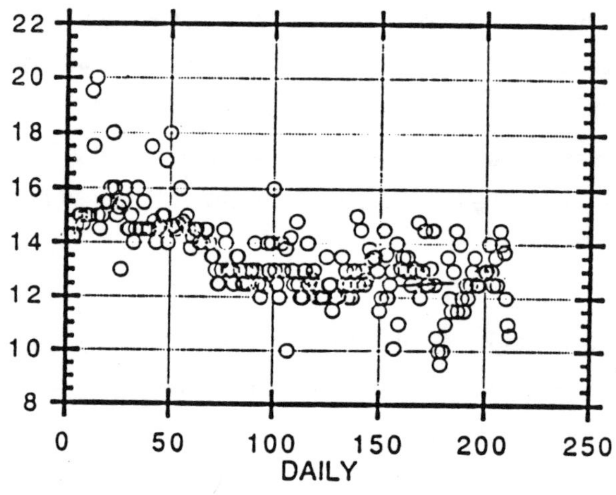

Figure 32. Influent Flow (mgd) Over Time for Santa Barbara's El Estero POTW

Table 13. Flows (mgd)

	10%	50%	90%
Goleta	4.4	6.0	6.7
Arcata	1.7	2.2	2.4
Santa Barbara	12.0	13.0	17.0
Los Angeles	157.0	178.0	185.0
Point Loma	180.0	190.0	200.0
Contra Costa	17.1	18.2	19.1

Table 14. Suspended Solids (mg/L)

	In			Out		
	10%	50%	90%	10%	50%	90%
Goleta	180	210	250			
Arcata	75	180	250	47	66	105
Santa Barbara (all)	200	400	800	5	8	15
Los Angeles	420	460	470	56	66	74
Point Loma	182	198	195	54	65	76
Contra Costa	100	140	190	70	90	125

Table 15. BOD

	In			Out		
	10%	50%	90%	10%	50%	90%
Goleta	220	250	420			
Arcata	-	-	-	-	-	-
Santa Barbara (all)	140	200	260	5	10	154
Los Angeles	340	360	380	105	108	115
Point Loma	-	-	-	-	-	-
Contra Costa	125	160	215	125	140	170

conserving technology at the end of the project. Data were calculated at 5-year intervals. Each scenario starts the community at the same level of conservation.

<u>Slow-Rate Implementation Scenario</u>. This scenario could simulate a community that has a low incentive to install conserving technology. By the end of the 20-year project period, 60 percent of the community still uses nonconserving devices (L0), and only 10 percent has installed L3 technology (Table 17).

<u>Medium-Rate Implementation Scenario</u>. This scenario models a community that is a little faster installing conservation technology.

<u>High-Rate Implementation Scenario</u>. This is a model of a community with strong incentive to reduce its water consumption. By the end of the project, this community has retrofitted every home that once had Level 0 (L0) technology (Table 18). Fifty percent of the people live in homes that utilize L2 devices. This represents a best-case scenario (Table 19).

<u>Findings</u>. Probably the most interesting result is the amount of reduction in wastewater flow for each plan. The plan that had a slow implementation rate showed an increase of about 0.3 mgd of wastewater flow per year, while the fast-rate plan reduced its flow at the 5-year point in the project, but the amount decreased was gained again by the year 10. From that point on, the daily wastewater flow remained more or less the same for the duration of the project. The medium-rate plan fell between the slow and the fast, in terms of wastewater flow. This implies that for a 25-year project, a fast-rate plan could stabilize the wastewater flow for a community of 100,000 persons who begin the project not conserving any water (see Figure 33).

The average monthly influent BOD and TSS of six communities in California varied significantly from 160 to 360 mg/L and 140 to 460 mg/L, respectively. This variation in influent concentration can account for significant differences in organic loadings and removal efficiencies in primary clarification. The 90 percentile BOD and SS for these communities varied from 215 to 420 mg/L and 190 to 800 mg/L, respectively.

The reduction in indoor water use concentrates the BOD and TSS concentration. These concentrating factors are not totally accounted for by the decrease of wastewater flows. It

Table 16. Summary Characteristics of POTWs in Study

	$Flow_{50}$ (mgd)	SS_{50} (mg/L)	BOD_{50} (mg/L)	SS_{50}/BOD_{50}
Goleta	6.0	210	250	0.84
Arcata	2.2	188	-	-
Santa Barbara (all)	13.0	400	200	2.00
Los Angeles	178.0	460	360	1.28
Point Loma	190.0	198	-	-
Contra Costa	17.1	140	160	0.875

Table 17. Percent of Population Using Technology—Slow-Rate Implementation Scenario

	Year 0	Year 5	Year 10	Year 15	Year 20
Level 0	100%	90%	80%	70%	60%
Level 1	0	5	10	20	20
Level 2	0	5	10	10	10
Level 3	0	0	0	0	10

Table 18. Percent of Population Using Technology—Medium Rate Implementation Scenario

	Year 0	Year 5	Year 10	Year 15	Year 20
Level 0	100%	80%	60%	45%	30%
Level 1	0	10	20	25	30
Level 2	0	5	10	15	20
Level 3	0	5	10	15	20

Table 19. Percent of Population Using Technology—High Rate Implementation Scenario

	Year 0	Year 5	Year 10	Year 15	Year 20
Level 0	100%	60%	40%	20%	0%
Level 1	0	20	20	15	10
Level 2	0	10	20	10	50
Level 3	0	10	20	25	40

appears that the TSS are solubilized to a greater extent in those communities where reduced wastewater flows were observed (Figure 34). It is suggested that the increased time of concentration allows for a greater solubilization of particulate and suspended BOD in domestic wastewater. No attempt was made in this phase of the study to determine the impact of solubilization of the TSS on the collection and treatment system. To some extent BOD could be removed within the collection system as the wastewater is exposed to a greater anaerobic contact period.

Significant increases in suspended solid removal occurs as the concentration of the influent suspended solids increases. This can best be shown with data from Central Contra Costa Sanitation District (Figure 35). The removal efficiency increases from 40 to 70 percent as the suspended solids in the influent increases from 100 to 200 mg/L. The mass removal rates increase also, as shown by the Point Loma data (Figure 36). At Point Loma the mass removal rate increases from 3.5×10^5 lbs/day to 4.2×10^5 lbs/day as the removal efficiency increases from 76 to 79 percent.

This analysis of the change in raw wastewater constituent as a result of reduced indoor water use represents a preliminary finding. The fact that drought conditions and water conservation strategies were occurring simultaneously makes it impossible to separate their distinct impacts. Obviously drought conditions force policies and programs of water conservation, which in turn collectively reduce wastewater flow. Exactly how much elasticity exists in indoor water use without structural changes is not known. The real test, though, for analyzing the impact of reduced inflow flows on wastewater collection and treatment will have to wait for normal rainfall conditions. Some general observations can be made under these sets of conditions which will help in the design of future information collection activities.

Based on information developed in this first study, the second phase should be directed at an in-depth study of selected wastewater collection and treatment system. The selected systems should meet the following criteria: 1) the system has a significant structural water conservation program, 2) plant process efficiency data is readily available, and 3) collection system information is readily available. The second phase of the study should include an in-depth analysis of operation and maintenance activities during the period of diminished flows

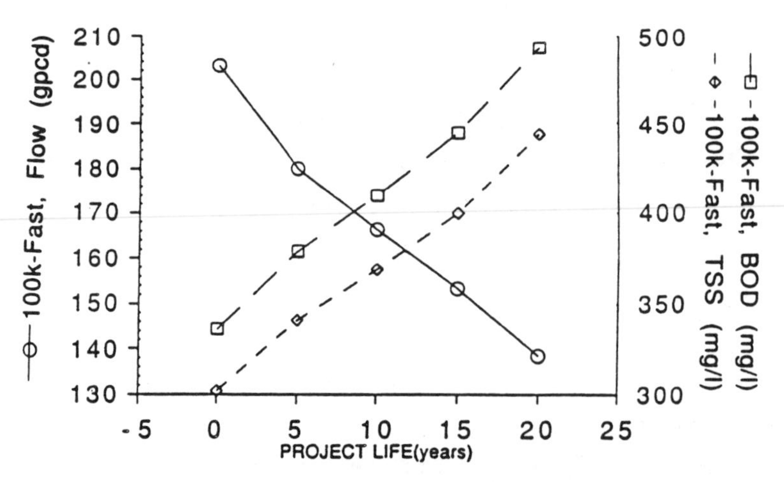

Figure 33. Predicted BOD, Suspended Solids, and Per Capita Wastewater Flow for High Rate Water Conservation Scenario

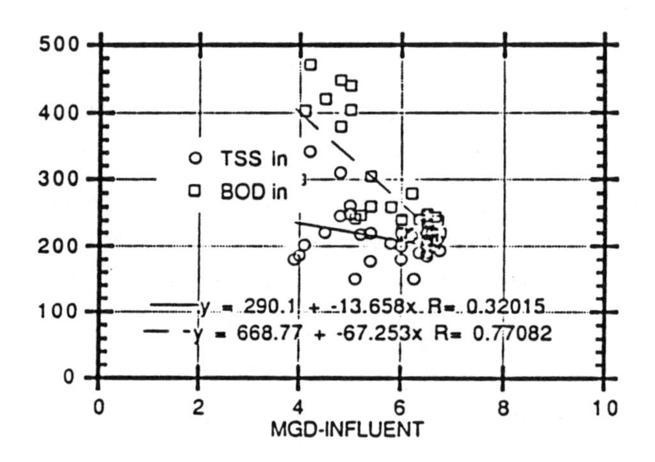

Figure 34. BOD and Suspended Solids vs. Influent Flow at Goleta POTW

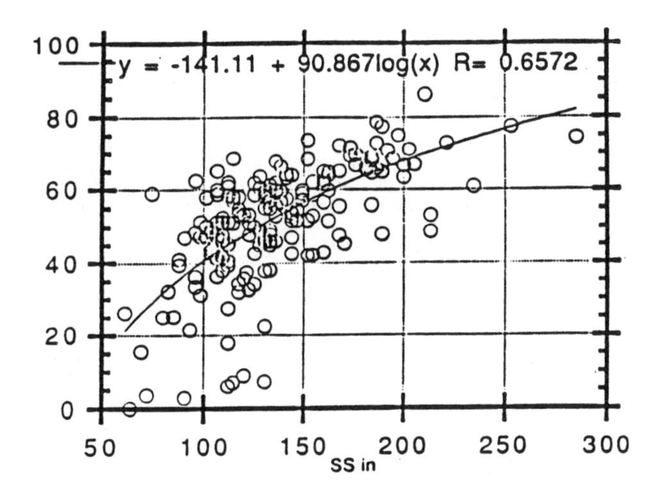

Figure 35. Percent Removal of Suspended Solids vs. Influent Suspended Solids for Central Contra Costa POTW

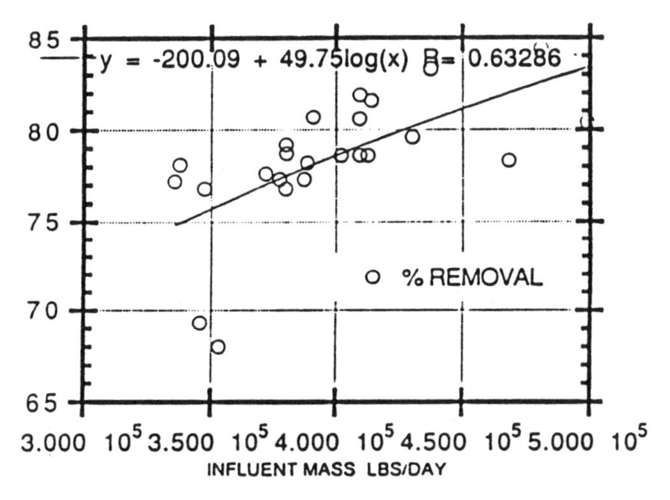

Figure 36. Percent Removal vs. Influent Suspended Solids Mass Loading (lbs/day) for Point Loma POTW

due to water conservation activity. These O&M requirements could then be compared with historic O&M requirements for a given type and size of wastewater collection/treatment system.

Summary. This study has incorporated operational data from six California POTWs, into a comparative analysis of influent wastewater characteristics. This study shows that indoor water conservation practices can significantly decrease the wastewater flows, thus increasing the influent BOD and suspended solids concentration, contrary to previous studies. This study shows the influence of the increase in time of concentration in the collection system as the ratio to BOD to suspended solids. It appears that the suspended material is solubilized, giving a greater BOD contribution in those systems with reduced wastewater flows.

Using a predictive model, it was shown that various scenarios of indoor water conservation could significantly alter wastewater flows over the planning period of wastewater treatment facility planning. For example, under the most restrictive scenario the wastewater flows would remain the same over a 20-year period. These scenarios combined levels of technology, percent of population involved, and percent of implementation. The BOD and suspended solids predicted from the model agree with operational data obtained from Goleta and Santa Barbara's water conservation efforts last year.

The efficiency of primary clarification increases with increased concentration of suspended solids. It appears, from preliminary analysis, that this is due to the combined effect of increased flux of suspended solids and increased retention times in the primary clarifier. The increase in primary clarifier solids can be considered a positive benefit if anaerobic digestion/cogeneration processes are involved.

The decrease in wastewater flows obviously will increase the retention time in various unit processes. This could be a significant factor in the case of Goleta, where average wastewater flows are reduced by 58 percent, increasing retention times in grit chambers, primary clarifiers, aeration units, secondary clarifiers, and chlorine contact basins.

The next phase of the study will look at in-plant impacts of indoor water conservation and collection systems. This phase will look in-depth at five POTWs, specifically Central Contra

Costa Sanitation District, East Bay Municipal Utility District, Goleta, and Los Angeles Sanitation District Joint Water Pollution Control Plant.

REFERENCES

1. Tchobanoglous, G. and E. D. Schroeder. 1985. Water Quality. Addison-Wesley Publishing Company.

2. Bailey, J. R., R. J. Benoit, J. L. Dodson, J. R. Robb, and H. Wallman. 1969. A Study of Flow Reduction and Treatment of Wastewater from Households. In General Dynamics, Electric Boat Division (ed.). Federal Water Quality Administration.

3. Consumers Union. 1990. How to Save Water, Consumer Reports. June (55):463-473.

4. Koyasako, J. S. 1980. The Effects of Water Conservation Induced Wastewater Flow Reduction—A Perspective. EPA-600/2-80-137.

5. Bohae, C. E. and R. A. Sierka. 1978. Effect of Water Conservation on Activated Sludge Kinetics, Journal of Water Pollution Control 30(10):2313-2326.

Fixed Film/Suspended Growth Secondary Treatment Systems

Arthur J. Condren, James A. Heldman and Bjorn Ruster

Introduction. The addition of inert media to support fixed film biomass growth in activated sludge aeration tankage offers the potential for cost-effective upgrading of municipal wastewater treatment systems. To gain a better perspective of the potential benefits of utilizing these high biomass systems, EPA undertook an investigation of plant performance at full-scale facilities employing several of these systems. Results and observations from a number of site visits to various European installations in mid-1988 are summarized in this paper.

Use of inert support media to serve as the locus for fixed film biomass growth is a relatively old concept. Many current approaches to high biomass systems employ a combination of fixed film and freely suspended biomass in the process. The suspended growth component concentration is controlled by adjusting the amount of MLSS wasted from the underflow of the system's secondary clarifiers. Since the fixed film biomass is retained in the system's aeration tankage, problems with solids-liquid separation in hydraulically overloaded secondary clarifiers can be addressed more easily with high biomass systems.

High biomass systems have gained a certain popularity in Europe. During the past few years, a number of investigations undertaken in the Federal Republic of Germany (FRG) have been reported (7-15). Among the advantages attributed to such systems have been improvements in nitrification performance, sludge settleability, and effluent quality (16,17).

Currently Available High Biomass Systems. At the present time, there appear to be at least six commercially available high biomass systems that can be incorporated into conventional aeration tanks. Linde AG of the FRG and Simon-Hartley of Great Britain (Ashbrook-Simon-Hartley in the USA) both use small, highly reticulated sponge pads as their inert support media. Bio-2-Sludge (FRG) and Smith & Loveless, Inc. (USA) use racks of synthetic trickling filter media to effect fixed film growth (see Figure 37). Ring Lace (Japan) employs a looped string material as the inert support media (Figure 38), and a Chinese firm uses tassels of a synthetic

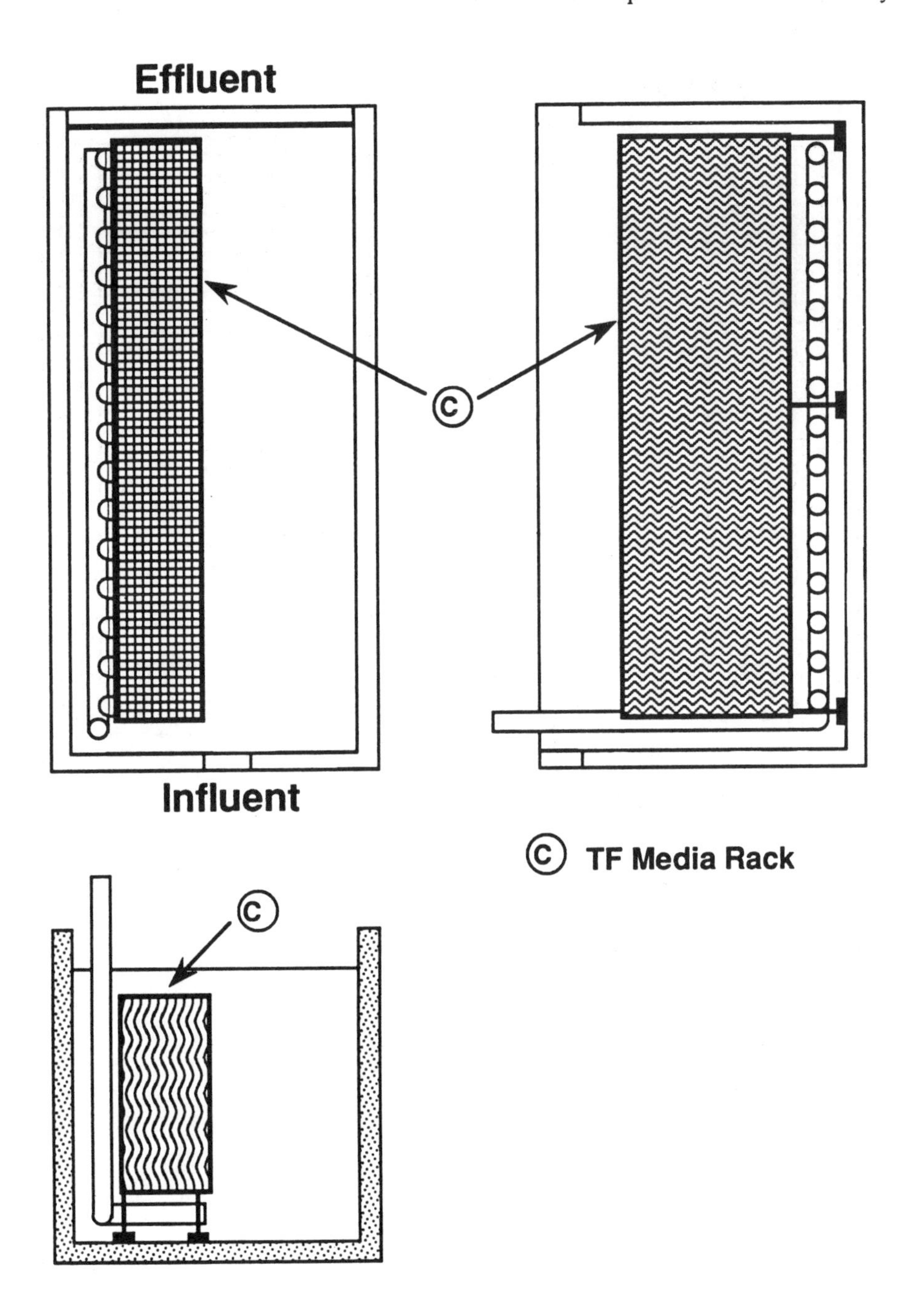

Figure 37. Plan and Section Views of a Bio-2-Sludge System

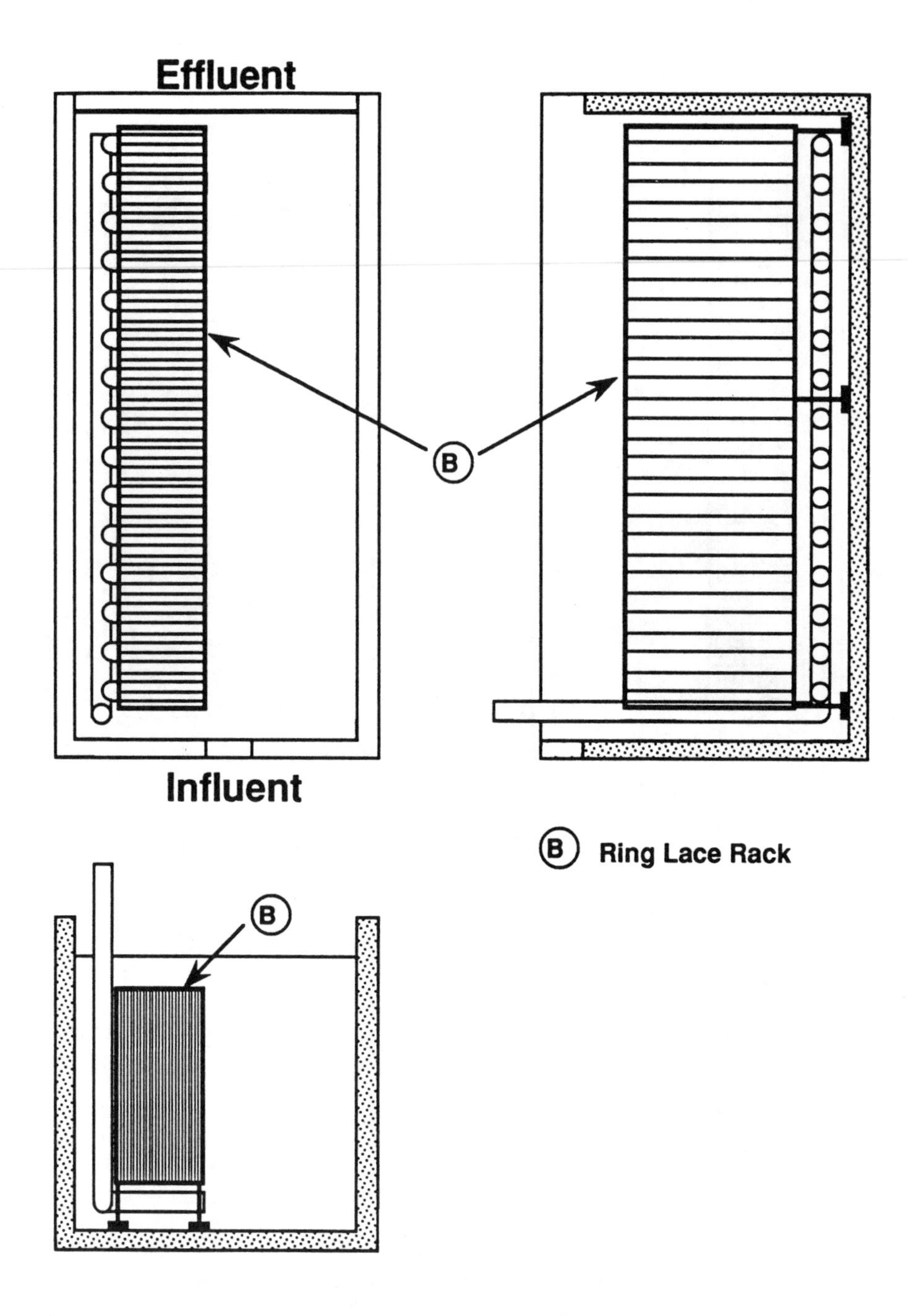

Figure 38. Plan and Section Views of a Ring Lace System

material attached to a string for fixed biomass attachment and growth. The latter two media are tied to racks which are placed in the plant's aeration tanks.

<u>Site Visits</u>. During mid-1988, a U.S. EPA evaluation team visited a number of full scale high biomass systems in the FRG. The purpose of the site visits was to view the systems in operation, collect operational and performance data, learn about system design details from the system manufacturers, and discuss process operational and maintenance concerns with the treatment plant staffs. Data and information on the various types of high biomass systems from these facilities are summarized below.

<u>Freising (Linde AG System)</u>. The Freising activated sludge plant was converted to a Linde AG high biomass system in 1984 after a series of pilot plant studies. The conversion employed a sponge volume equal to 20 percent of the aeration tank volume. It appears that there were three primary reasons for the conversion: frequently occurring poor sludge settleability, limited space at the plant site, and higher costs of alternative technologies.

Operational and performance data were collected for a short period of time before and after plant conversion. Before conversion, the plant could only maintain a MLSS concentration of about 2,600 mg/L. Following conversion, a much higher MLSS concentration could be maintained, which dramatically lowered the F/M of the system and greatly improved secondary sludge settleability.

As a follow-up to this historic information, data from 1987 were collected and analyzed. In 1987, the Freising plant was operated at 78 percent of its hydraulic capacity and 80 percent of its BOD_5 capacity. Instantaneous influent pH, because of industrial discharges, ranged from 6.2 to 12.0; effluent pH ranged from 6.8 to 7.5. An average of 65 percent nitrification was achieved over a wastewater temperature range of 10 to 17 C, even though the dissolved oxygen concentration averaged only 1.7 mg/L.

<u>Munich (Linde AG System)</u>. Munich's Grosslappen plant was retrofitted with the Linde AG high biomass system to allow for additional treatment capacity while a parallel activated sludge plant of equal capacity was being constructed. Sponge media were installed to allow for full

scale evaluation of the process and for possible inclusion in the new treatment plant design. The existing Munich plant has three banks of aeration tanks, and during the site visit, two of the banks had varying quantities of sponges in place.

Points of interest from the Munich data are: 1) a lowering of the system's F/M by the presence of the fixed film biomass, 2) an increase in aeration rate to address the demand of additional biomass in the system, and 3) improved effluent quality.

Olching (Ring Lace System). During the spring of 1988, the Olching plant was operating at 85 percent of its design hydraulic capacity of 35,000 m^3/day and 83 percent of its design organic loading, which was based on a population equivalent of 240,000. In late 1987, conversion of the plant to a high biomass system began by adding the installation of a new 4,000 m^3 denitrification basin and the addition of 252,000 m of Ring Lace material to each of four 2,070 m^3 aeration tanks. The lead denitrification basin, which contains no Ring Lace and had not yet been placed in operation at the time of the site visit, is equipped with paddle mixers and also contains aeration equipment for nitrification if necessary. A portion of the mixed liquor from the Ring Lace basins will eventually be recycled to the lead denitrification basin. At the time of the site visit, influent flow was sent directly to the aeration basins containing the Ring Lace material.

The Ring Lace material was strung on racks, with the individual strands being separated by approximately 50 mm. Spacing between the racks, which occupied 31 percent of the aeration tank volume, was about 60 mm. A tensioning system was built into the racks in case the Ring Lace material elongates with time. In addition to the above, the equipment supplier had to provide a 10-year process performance and equipment guarantee. Required process performance was based on the plant's discharge requirements of 15 mg/L BOD_5, 20 mg/L TSS and 10 mg/L NH_4-N, the latter equating to an approximate 75 percent level of nitrification.

Because the Olching Ring Lace plant had been in operation for less than one year and the required performance test had not been completed, no operational or performance data were released to the U.S. EPA evaluation team. However, the following general observations on the Ring Lace system were communicated from the treatment plant staff.

- Before conversion to the high biomass system, the maximum operational MLSS concentration that could be achieved was about 1,500 mg/L, which resulted in the plant operating at a F/M of 0.6 to 0.70 kg BOD_5/kg MLSS-day. At this loading rate the MLSS were gray in color and had very poor settling characteristics. In addition, no substantial nitrification could be achieved.

- After conversion, the suspended biomass varied from 3,500 to 4,500 mg/L. Fixed biomass on the Ring Lace material was estimated at 6.5 g/m, which is equivalent to an additional 790 mg/L of MLSS. This increase resulted in a MLSS with a rich brown color and greatly improved settling characteristics. Operating F/M of the system decreased to about 0.2 kg/kg-day and desired levels of nitrification were being achieved.

- Other high biomass systems were evaluated at the pilot plant level by the Olching treatment plant staff and they all yielded effluent qualities equivalent to that realized by the Ring Lace system. However, the Linde AG system was not selected because the staff felt that this system might require additional electrical power for proper operation, the sponge cubes would be subject to wear by abrasion and, upon extended levels of mineralization, the sponge cubes might have a tendency to settle in the aeration tanks. The Bio-2-Sludge process was not selected because the plant staff felt that the synthetic trickling filter media might be subject to plugging by biomass and/or large particulates, either of which might lead to anaerobic zones in the media. Such zones, if they developed, were felt to potentially limit the nitrification process.

Schomberg (Bio-2-Sludge System). The Schomberg wastewater treatment plant was recently expanded in throughput capacity and simultaneously converted to a Bio-2-Sludge high biomass system. Both aeration tank volume and final clarifier surface area were more than doubled, and denitrification tankage was also added. Tropac media at 120 m^2/m^3 was added to 25 percent of the aeration tank volume. In the spring of 1988, the plant was operating at 20 percent of its design hydraulic capacity and 57 percent of its design BOD_5 capacity.

Prior to conversion, the plant was slightly overloaded, and could not meet discharge permit requirements primarily because of very poor settling MLSS. The poor settling characteristics would allow the maintenance of only 1,000 to 1,500 mg/L MLSS in the system's aeration tanks in spite of the very high RAS pumping. After conversion, overall plant performance greatly increased.

Installation of the Bio-2-Sludge system at Schomberg along with the other plant modifications greatly enhanced MLSS settleability as evidenced by the SVI of 82 mg/L after conversion. Although the RAS rate at this plant could have been adjusted to about 50 percent, a return rate of 100 percent was used. It appears that this is a common design/operational practice at smaller treatment plants in certain areas of Germany. It is interesting to note that this plant, in addition to achieving an effluent containing 5 mg/L BOD_5, realized nearly complete nitrification.

Calw-Hirsau (Bio-2-Sludge System). In 1986, the Calw-Hirsau facility was upgraded from a population equivalent design capacity of 23,000 to 53,000. Major changes included discontinuing operation as a LURGI plant (chemical addition to the secondary clarifers), addition of Tropac media (167 m^2/m^3) to 26 percent of the existing 1,617 m^3 aeration tank, and an increase in secondary clarifier capacity from 230 m^2 to 1,267 m^2 of surface area. Prior to plant conversion, there were reported problems with sludge bulking, but no data were available to quantify severity of the situation. Currently, process stability is reported to be good and maintenance has posed no problems.

System Economics. It is interesting to compare the costs of upgrading an activated sludge facility by high biomass and conventional approaches. Consider the hypothetical situation summarized in Table 20, in which a 18,930 m^3/day (5 mgd) plant is operated at a F/M of 0.33 kg BOD_5/kg MLVSS·day and a secondary clarifier solids loading rate of 3.83 kg/m^2·hr. Because of operational instability and effluent quality excursions, plant upgrading will be undertaken. There are basically three upgrading options available:

- Increase the volume of the plant's aeration tanks and operate the facility at a lesser MLVSS concentration.
- Increase the surface area of the plant's secondary clarifiers and operate the facility at a higher MLVSS concentration.
- Use the plant's existing tankage, but convert the facility to one incorporating a high biomass system.

Conventional upgrading requirements were determined to be a F/M of 0.20 kg BOD_5/kg MLVSS·d and a solids loading on the secondary clarifiers of 2.85 kg/m^2·hr (14 lb/d·ft^2). One approach, summarized in Table 20, would be to add aeration tankage (Case A); while an alternative approach would be to add secondary clarifiers (Case B).

Table 20. Conventional Upgrading Requirements

Parameter	Existing Plant	Upgrading Option A	Upgrading Option B
Secondary System Influent			
Flow, m3/day	18,930	18,930	18,930
BOD_5, mg/L	154	154	154
TSS in Aeration Tanks			
Suspended, mg/L	2,500	2,044	4,107
MLVSS/MLSS, %	75	75	75
Aeration Tanks			
Volume, m3	4,733	9,508	4,733
Det'n Time, hr	6.0	12.1	6.0
F/M, kg/kg-day	0.33	0.20	0.20
SVI, mL/g	133	133	133
Secondary Clarifiers			
Area, m2	774	774	2,513
SOR, m/hr	1.02	1.02	0.31
SLR, kg/m2-hr	3.83	2.85	2.85
Return Activated Sludge			
Concentration, mg/L	7,500	7,500	7,500
R/Q, %	50	37	121

Option A - Increase aeration tank volume/decrease MLVSS concentration
Option B - Increase secondary clarifier area/increase MLVSS concentration

Estimated total installed cost (design through start-up services) for Option A is approximately $1,820,000, while that for Option B is about $2,355,000. Option A facilitiesinclude the new aeration tanks and associated air headers and diffusers, and a new splitter box for equal flow distribution to all of the plant's aeration tanks. Option B facilities include the new secondary clarifiers and associated return and waste activated sludge pumps and piping, and a new splitter box for equal flow distribution to all of the plant's secondary clarifiers. The costs expressed per unit of original aeration tank volume are $385/m^3 of original tank volume for Option A and $498/m^3 of original volume for Option B. In all cases, the unit volume costs exceed the typical costs reported for the high biomass systems evaluated (e.g.,

Linpor at \$250 to \$350/m^3 of aeration tank volume; Ring Lace at \$300 to \$350/m^3 of aeration tank volume; and Bio-2-Sludge at \$300 to 350/m^3 of media installed).

Unfortunately there is no general design approach for sizing high biomass systems so that their expected performance would be equivalent to the conventional activated sludge upgrading alternatives evaluated in this design example. Based on available information, pilot scale studies are necessary to properly size any of the high biomass systems.

Data gathered during the site visits which summarize the amount of immobilized biomass in the fixed media systems are shown in Table 21. For both Linpor and Bio-2-Sludge there are three-fold variations in the amount of immobilized biomass per unit of media. Furthermore, it is unknown how immobilized biomass concentrations would normally vary as a function of loading (F/M) or of other system parameters such as aeration intensity. Mass transfer considerations are important in the evaluation of fixed media systems, and there may be considerable difference in the percentage of active attached biomass among the various fixed media systems as well as unit substrate removal rates (kg BOD/kg attached VSS) in combined fixed/suspended systems at the same overall volumetric and total mass loading.

Table 21. Observed Inert Support Media Biomass Values

System	Location	Immobilized Biomass	Overall System F/M
Ring Lace	Olching	6.5 g/m	0.17
Linpor	Freising	18.7 g/L of pad[a]	0.19
	Munich	12.0 g/L of pad[a]	0.57
	Munich	6.1 g/L of pad[a]	0.46
Bio-2-Sludge	Calw-Hirsau	5775 g/m^3 of media	0.15
	Schomberg	1770 g/m^3 of media	0.08

[a]At a theoretical pad density of 100% by volume

If the 5,775 g/m^3 of biomass observed for the Bio-2-Sludge system at Calw-Hirsau were achieved in the hypothetical upgrading problem (Table 20) by adding trickling filter media to the aeration tank, only 1,691 m^3 of media would be required to give an overall system F:M of 0.20 with 2,044 mg/L of suspended MLSS; this provides the desired solids loading on the secondary clarifier. Based on the German data, the installed cost for such a system would only be $390,000. An independent estimate undertaken by a U.S. manufacturer of wastewater treatment systems for this trickling filter media system would be about $850,000. These costs are 19 to 47 percent of the lowest cost conventional alternative evaluated.

A similar analysis can also undertaken for the Linpor and Ring Lace Systems. At 18.7 g/L of pads, a Linpor system would require about 11% pads by volume to achieve a total system biomass loading of 0.2 (suspended MLSS of 2,044 mg/L). This is at the low end of Linpor pad densities, suggesting that an installed cost of $250/m^3 may be applicable. This is equivalent to a total cost of $1,182,000, or 65 percent of the lowest cost conventional alternative. The reader must bear in mind that these values are based on FRG material, construction, and labor costs, which are not equivalent to those in the United States. The low biomass density on the Ring Lace system shown in Table 21 equates to an installed Ring Lace density of 317 m/m^3 of aeration tankage to achieve an equivalent immobilized biomass of 2,063 mg/L (suspended MLSS of 2,044 mg/L). At $2.65/m of Ring Lace, this equals a total cost of nearly $4,000,000.

Given the significant differences in mass transfer characteristics and the percentage of aerobic biomass likely to exist among the various inert media systems, comparison on the sole basis of mass of immobilized biomass is tenuous at best. Nonetheless, this simple type of analysis does suggest that at least some of the high biomass systems are potentially cost-effective upgrading approaches at overloaded conventional activated sludge plants. In addition, the inclusion of inert support media may allow for similar cost-effective achievement of nitrification at facilities originally designed to accomplish only carbonaceous BOD_5 removal.

<u>Summary</u>. A number of high biomass systems have been installed at wastewater treatment plants throughout the Federal Republic of Germany. These installations were designed to effect improved effluent quality, and it appears that this goal is being realized.

Reasons for selecting high biomass systems over construction of additional aeration tanks and clarifiers (or other secondary treatment processes) include reduced space requirements, increased process stability, and capital/operating cost savings.

High biomass systems call for installation of supplemental equipment over that contained in a conventional activated sludge plant. More installed equipment generally implies more maintenance, and, to some extent, this is true for some of the systems. In addition, the presence of both suspended and fixed biomass forms and higher biomass concentrations may require a certain level of additional operator time to achieve optimum system performance.

The presence of inert support media and higher biomass concentrations in these systems can increase overall power consumption. To achieve desired mixing patterns in retrofitted aeration tanks, power input may have to be increased. Also, the presence of additional biomass increases system oxygen requirements which, in turn, requires additional power input. In addition, high biomass systems generally yield higher levels of nitrification, which also can affect overall power consumption. Such factors should be addressed when analyzing operating costs.

Certain system design limitations have been identified in the past, but many of these have been corrected by the system manufacturers and/or operators. For example, influent hydraulic surges at Linde AG plants have caused partial blinding of the retention screens by the sponge media. This has been partially corrected by increasing the pumping rate of the sponge return system. Also, abrasion of the sponge media was stated to result in losses of less than 5 percent per year, but there does not appear to be a corrective action for this system limitation beyond adding new sponges as required. In Ring Lace systems, a question on the extent of media stretching over time persists. Construction of self-tensioning media racks may address this potential limitation. Plugging of the synthetic trickling filter media in Bio-2-Sludge systems has been a concern of certain individuals. It appears that periodic use of an additional air blower induces additional sloughing of the fixed film biomass, thus preventing this potential problem. There were other problems communicated to the EPA evaluation team, but these were of minor importance because solutions had, for the most part, already been developed and field tested.

The capacity for increased throughput realized by conversion to a high biomass system cannot be assessed from available data. However, it appears that conversion can result in at least the doubling of the biomass concentration in a system and for the reduction by at least one-third of the required RAS pumping rate, compared to a conventional activated sludge plant. These two factors alone indicate the potential for increased throughput capacity. A preliminary economic analysis suggests that high biomass systems may be more cost effective for plant upgrading than conventional approaches.

The actual capacity increase that can be realized by conversion to a high biomass system should be established from the conduct of pilot plant studies. Representatives for both Bio-2-Sludge and Linde AG stated that to obtain accurate design and operational information, a pilot plant should have an aeration tank with a volume of at least 100 m^3.

REFERENCES

1. Peters, P. W. and J.E. Alleman. 1982. "The History of Fixed-Film Wastewater Treatment Systems," Proc. First Internat. Conf. on Fixed-Film Biological Processes, Vol. 1, p. 60.

2. Wilford, J. and T.P. Conlon. 1957. "Contact Aeration Sewage Treatment Plants in New Jersey," Sewage and Industrial Wastes, 29: 845.

3. Huang, C.S. 1982. "The Air Force Experience in Fixed-Film Biological Processes," Proc. First Internat. Conf. of Fixed-Film Biological Processes, Vol. 3, p. 1777.

4. Boyle, W. C. and A.T. Wallace. 1986. "Status of Porous Biomass Support Systems for Wastewater Treatment: An I/A Technology Assessment," EPA/600/S2-86/019, EPA, Cincinnati, OH.

5. Heidman, J.A., R.C. Brenner, and H.J. Shah. 1988. "Pilot-Plant Evaluation of Porous Biomass Supports," Jour. Env. Eng. Div., ASCE, 114: 1077.

6. Arora, M.L. and M.B. Umphres. 1987. "Evaluation of Activated Biofiltration and Activated Biofiltration/Activated Sludge Technologies," Jour. Water Poll. Control Fed., 59: 183.

7. Hegemann, W. and A. Wildmoser. 1986. "Sanierung einer Belebungsanlage durch den Einsatz von schwimmenden Aufwuchskorpern zur Biomassenanreichung," gwf-wasser/abwasser, (9):415-421.

8. Lassmann, E. and H. Reimann. 1987. "Der Einsatz von offenporigem Schaumstoff als Tragermaterial bei der biologischen Abwasser-Reinigung," Chemie-Ingenieur-Technik, (59):2:132-134.

9. Lang, H. 1981. "Nitrifikation in biologischen Klarstufen mit Hilfe des 'Bio-2-Schlmann-Verfahrens'," Wasserwirtschaft, 71(6):166-169.

10. Eberhardt, H., O. Klee, and W. Weber. 1984. "Leistungssteigerung einer uberelasteten Belebungsanlage durch Einbau submerser Festkorper," Wasserwirtschaft, 74(2):47-53.

11. Schlegel, S. 1984. "Ergebnisse von Versuchen einer aus Tropfkorper mit nachgeschalteter Belebung bestehenden Anlage im Vergleich zur einstufigen Belebungsanlage," Korrespondenz Abwasser, 31(3):252-253.

12. Schlegel, S. 1987. "Uberden Einsatz von getauchten Festbettkorpern bei der biologischen Abwasserreinigung," Chemie-Ingenieur-Technik, 59(3):252-253.

13. Schlegel, S. 1988. "Der Einsatz von getauchten Festbettkorpern zur Nitrifikation," Korrespondenz Abwasser, 35(2):120-126.

14. Schlegel, S. 1988. "The use of Submerged Biological Filters for Nitrification," Wat. Sci. Techn., 20(4/5):177-187.

15. Scherb, K. 1987. "Abwasserreinigung nach dem Ring-Lace-Verfahren (Bioringschnur-Verfahren)," Muchener Beitrage zur Abwasser, Fischereiund Flussbiologie, Band 41: Stand der Technik bei der Elimination umweltrelevanter Abwasserinhaltsstoffe, pp. 212-229, R. Oldenbourg, Muchen.

16. Wanner, J., K. Kucman, and P. Grau. 1988. "Activated Sludge Process Combined with Biofilm Cultivation," Wat. Res., 22(2):207-215.

17. Rogalla, F. et al. 1988. "Fixed Biomass to Upgrade Activated Sludge," Presented 61st Annual WPCF Conference, Dallas, Texas, October.

Chemical Phosphorus Removal in Lagoons

Charles Pycha

U.S. Environmental Protection Agency
Chicago, Illinois

<u>**Introduction**</u>. Phosphorus has been a contaminant of concern to EPA's Region 5 for many years primarily because of the Great Lakes as well as numerous other smaller lakes and impoundments. The basic chemistry of biological and chemical removal of phosphorus is covered quite adequately in the EPA design manuals on phosphorus removal. What these manuals do not address is removal of phosphorus from lagoon or pond systems. The technology of removing phosphorus from ponds, or P-Ponds, is not really new but has not been used to a great extent. The EPA Design Manual on Municipal Wastewater Stabilization Ponds does address this technology. The purpose of this report is to take a more detailed look at how P-Ponds are utilized in Region 5, principally in Michigan and Minnesota.

<u>**Background**</u>. In the early 1970s, the Ontario Ministry of the Environment initiated several research projects on nutrient control in sewage lagoons. These reports provide the baseline information upon which most applications of this technology were designed around. These research projects addressed continuous as well as seasonal discharge lagoons. They covered the addition of ferric chloride, alum, and lime at various dosages. Based on these studies, Ontario decided to design and operate full-scale municipal seasonal discharge spring and fall lagoons using alum to precipitate the phosphorus. Provincial personnel provide the manpower for the chemical addition and discharge for over 20 lagoon systems. Their methodology, which is akin to a contract operation, varies somewhat from Region 5.

<u>**Region 5 Experience**</u>. Ontario uses a team of four or five people and several boats. Each boat is equipped with a tank for alum with a pump to inject the alum into the propwash at the stern of the boat. Their stabilization ponds are usually 10 acres or more. They will typically have a tank wagon with alum driven directly to the lagoon site, along with 3 boats 14 to 16 ft in length with 50 or 60 HP motors that they operate simultaneously. They can almost continuously load alum to the boats and apply it to the total lagoon cell in a couple of hours. The pond will be given a day to settle and then the level will be lowered 3 to 4 ft over the period of 5 to 10 days.

Using these procedures, the operating team can service a large number of lagoons in one geographic area simultaneously. Ontario's operating approach appears to generally stay well above the stoichiometric dosage rate, which usually ensures low effluent total phosphorus residual levels often staying consistently below 0.5 mg/L.

Minnesota, the legendary land of 10,000 lakes, with hundreds of municipal lagoon systems, has about a dozen that practice chemical addition (alum) directly to the final cells via motorboat to achieve reduction of total phosphorus to meet a 1.0 mg/L effluent standard. These lagoon systems range from a design flow of 25,000 gallons per day with a polishing lagoon as small as 1.5 acres to a design flow of 0.7 mg/L with a polishing pond size of up to 20 acres. Alum is applied to the lagoon surface with boats ranging from a 12-ft aluminum boat with a 5 HP outboard motor to a 17-ft pontoon boat with a 500 gal onboard storage tank and twin 50 HP motors for power. At most facilities liquid alum is used; bags of powdered alum are used at some of the smaller facilities because they are more convenient to store.

The majority of the facilities will inject the alum into the outboard motors propwash, following the Ontario example. Several of the smaller facilities have outrigger arms to pump the alum onto the lagoon surface several feet to either side of the boat. This method, though ensuring full surface coverage, would not appear to as thoroughly mix the alum with the pond water, as would adding it to the propwash. It is successful in meeting a 1.0 mg/L effluent limit possibly because it appears that the initial mixing plus the added mixing from the distribution boats' cross-hatching pattern adequately precipitates the phosphorus. Influent phosphorus values ranged as low as 3.0 mg/L, which is significantly lower than historic published values, possibly due to the widespread marketing of nonphosphate detergents.

Michigan has taken a slightly different approach to phosphorus removal from lagoon systems. There are some 26 lagoon facilities with phosphorus limits that fall into 3 general types: 1) seasonal discharge, 2) continuous discharge with chemicals being added to the polishing pond, and 3) continuous discharge with clarifiers following the pond system. Sizes of these facilities range from 0.25 to 7.5 mgd. There is a wide combination of chemicals that are used including ferric chloride or alum at equal numbers of facilities. One facility uses both ferric chloride and alum, while several facilities use a polymer in conjunction with either metal or salt. A few of these facilities have been operating for as long as 20 years. Most of these facilities add the chemicals between lagoon cells, often using a mixing chamber between the cells or between the lagoons and the final clarifier. The chemicals are added continuously or more specifically whenever wastewater is being transferred into the polishing pond or clarifier.

Discharge procedures for these systems vary from spring and fall discharge periods of several weeks, to 24 or 8 hours per day on a continuous basis throughout the year. Additionally, permit effluent limits, based upon total phosphorus, which are typically written in the standard 1.0 mg/L based on a 30-day average with a resultant pounds per day calculated based upon design flow, also may include pounds per day daily maximum value and pounds per day based upon a 30-day average without a concentration limit.

Conclusion. This causes a problem when one tries to draw a general conclusion regarding Region 5 experience with P-Ponds. The conclusion is that the technology of adding chemicals to precipitate phosphorus in lagoons is effective but there are problems. Only two of the 30 facilities reviewed as part of this report were in significant noncompliance, but there were many minor excursions in excess of the permitted phosphorus limits at other facilities. Problems identified included those typical of lagoon systems, namely mixing of the lagoon contents by wind and waves during discharge as well as the expected algal blooms. Other typical problems with the storage, pumping, and mixing of the chemicals were to be expected. Only one facility identified a problem with resolubilization of the phosphorus in the bottom sediment, and this was related to a change in the pond pH due to algal blooms, which can affect the reaction with the ferric chloride used at this facility. Several of the seasonally discharged P-Ponds based the initial chemical dose on past experience rather than by calculating the required dose based upon the initial phosphorus concentration in the ponds. This could be a result of a tendency to minimize operating costs (purchasing of excessive chemicals) and may be the cause of some of the minor permit excursions that were documented. This varies from the Ontario experiences where they prefer to add a heavier dose of chemical to ensure permit compliance as well as reaping a secondary benefit of discharges of less phosphorus (and BOD and SS that is also precipitated) to the receiving waters.

On the positive side of these experiences is the fact that the seasonal discharge lagoons require a minimum of operator attention. Chemical addition on a batch basis is easily calculated and easily applied through an influent structure or via motorboat. The systems can and have regularly achieved effluent total phosphorus limits of 1 mg/L or less under a wide variety of lagoon configurations, climatic conditions, and a wide range of design flow rates (0.025 to 7.5 mgd). As with any treatment system, P-Ponds depend upon operator knowledge and attention. Chemical precipitation on BOD and SS is also of benefit to pond systems because of the wide seasonal variability of the organic life within the pond.

APPENDICES

Appendix A—Agenda

U.S. Environmental Protection Agency
1991 Wastewater Technology Forum Agenda

June 5–7, 1991
Portland Hilton
Portland, Oregon

TUESDAY, JUNE 4, 1991

5:00 p.m. - 7:00 p.m. Poster Session (Parlors A, B & C)

6:00 p.m. - 9:00 p.m. Early Registration

WEDNESDAY, JUNE 5, 1991

7:30 a.m. Registration

8:30 a.m. Introduction

- Opening Remarks
  Wendy Bell, U.S. EPA, Office of Wastewater Enforcement and Compliance (OWEC), Washington, DC

- Welcome
  Bob Burd, U.S. EPA, Region 10

- Keynote Address
  Mike Cook, U.S. EPA, OWEC, Washington, DC

- Some Thoughts on Wastewater Technology in the 90s
  Bob Lee, U.S. EPA, OWEC, Washington, DC

9:30 a.m. Update on EPA's Sludge Policy and New Sewage Sludge Regulations
Bob Bastian, U.S. EPA, Washington, DC

10:15 a.m. Break

WEDNESDAY, JUNE 5, 1991
(continued)

10:30 a.m. — Land Treatment
Moderator: Bryan Yim, U.S. EPA, Region 10

- Considerations for Overland Flow Diagnostic Evaluations
  Tom Wooters/Chris King, Crowder College, Neosho, MO

- Best Available Technology for Design and Siting for Land Application of Wastewater on the Rathdrum Prairie--Kootenai County, Idaho
  John Sutherland, State of Idaho, Division of Environmental Quality, Coeur D'Alene, ID

11:45 a.m. — Lunch with Speaker
Speaker: Skeet Arasmith, Arasmith Consulting Resources, Albany, OR

1:15 p.m. — Sand and Gravel Filters
Moderator: Steve Hogye, U.S. EPA, Washington, DC

- Sand Filters: State of the Art
  Harold Ball, Orenco Systems, Inc., Roseburg, OR

- Tennessee Experience with Recirculating Sand Filters: Wastewater Treatment Systems for Small Flows
  Steve Fishel, Tennessee Division of Water Pollution Control, Nashville, TN

- Recirculating Gravel Filters in Oregon
  Jim Van Domelen, Oregon Department of Environmental Quality, Portland, OR

3:00 p.m. — Break

3:15 p.m. — Operations and Maintenance
Moderator: Lam Lim, U.S. EPA, Washington, DC

- Assessment of O & M Requirements for Ultraviolet Disinfection Systems
  Karl Scheible, HydroQual, Inc., Mahwah, NJ

- Trickling Filter O&M Issues
  Russell Martin, U.S. EPA, Region 5

- Update on the Microbial Rock Plant Filter
  Ancil Jones, U.S. EPA, Region 6

WEDNESDAY, JUNE 5, 1991
(continued)

5:15 p.m.	Adjourn
5:15 p.m. - 7:00 p.m.	Poster Session (Parlors A, B & C)

THURSDAY, JUNE 6, 1991

8:00 a.m.	Biological Nutrient Removal Moderator: Atal Eralp, U.S. EPA, Washington, DC
•	Meeting More Stringent Standards Using BNR Glen Daigger, CH2M Hill, Denver, CO
•	Operation of BNR systems at two Oregon POTWs Gordon Nicholson, CH2M Hill, Bellevue, WA
•	Summary of Patented and Public Biological Phosphorous Removal Systems William Boyle, University of Wisconsin, Madison, WI
9:45 a.m.	Break
10:00 a.m.	Sludge Moderator: John Walker, U.S. EPA, Washington, DC
•	Case Study Evaluation of Alkaline Stabilization Processes Lori Stone, Engineering-Science, Inc., Fairfax, VA
•	Controlling Sludge Composting Odors William Horst, City of Lancaster, PA
•	Co-composting Dale Cap, Southwest Suburban Sewer District, Seattle, WA
11:45 a.m.	Lunch - On Your Own
1:00 p.m.	Field Trip to two Portland Wastewater Treatment Plants: 1. Tri-City Water Pollution Control Plant Highlighting the Anoxic Selector Activated Sludge System for Nitrogen Removal 2. Columbia Boulevard Wastewater Treatment Plant Highlighting the In-Vessel Composting System

5:00 p.m.	Arrive Back at Hotel

FRIDAY, JUNE 7, 1991

8:00 a.m.	Stormwater Moderator: Jim Kreissl, U.S. EPA, Center for Environmental Research Information (CERI), Cincinnati, OH
•	Development of CSO Regulations in Washington State Ed O'Brien, Washington Department of Ecology, Olympia, WA
•	Cost of CSO Controls Atal Eralp
•	Stormwater Control for Puget Sound Peter Birch, Washington Department of Ecology, Olympia, WA
9:45 a.m.	Disinfection Moderator: Jim Kreissl
•	Total Residual Chlorine: Toxicological Effects and Fate in Freshwater Streams in New York State Gary Neuderfer, New York Department of Environmental Conservation, Avon, NY
•	EPA Disinfection Policy and Guidance Update Bob Bastian
10:30 a.m.	Break
10:45 a.m.	Constructed Wetlands Moderator: Bob Bastian
•	Arcata, CA Bob Gearheart, Humboldt State University, Arcata, CA
•	Constructed Wetlands Experience in the Southeast Bob Freeman, Cobb County Water System, Marietta, GA
12:00 p.m.	Lunch - On Your Own
1:15 p.m.	Municipal Water Use Efficiency Moderator: Bob Bastian
•	How Efficient Water Use Can Help Communities Meet Environmental Objectives Steve Hogye

FRIDAY, JUNE 7, 1991
(continued)

•	Effects on POTWs Bob Gearheart
2:15 p.m.	High Biomass in Europe Art Condren, James Montgomery Consulting Engineers, Pasadena, CA
3:00 p.m.	Chemical Phosphorus Removal in Lagoons Chuck Pycha, U.S. EPA, Region 5
3:45 p.m.	End of Forum

Appendix B—Speaker List

U.S. Environmental Protection Agency
1991 Wastewater Technology Forum Speaker List

June 5-7, 1991
The Portland Hilton
Portland Oregon

Skeet Arasmith
Arasmith Consulting Resources
1298 Elm Street, SW
Albany, OR 97321
503-928-5211

Harold Ball
Orenco Systems, Inc.
2826 Colonial Road
Roseburg, OR 97470
503-673-0165

Robert Bastian
Office of Wastewater
Enforcement & Compliance
U.S. Environmental Protection Agency
401 M Street, SW (WH-547)
Washington, DC 20460
202-382-7378

Wendy Bell
Office of Wastewater
Enforcement & Compliance
U.S. Environmental Protection Agency
401 M Street, SW (WH-547)
Washington, DC 20460
202-382-7292

Peter Birch
Water Quality Program
Washington Department of Ecology
Olympia, WA 98504
206-438-7076

William Boyle
Department of Civil &
Environmental Engineering
University of Wisconsin
3230 Engineering Building
Madison, WI 53706
608-262-1777

Bob Burd
Water Division
U.S. Environmental Protection Agency
1200 Sixth Avenue
Seattle, WA 98101
206-553-1014

Dale Cap
Southwest Suburban Sewer District
1015 Southwest 174th Street
Seattle, WA 98166
206-242-7907

Arthur Condren
James Montgomery Consulting Engineers, Inc.
250 North Madison Avenue - P.O. Box 7009
Pasadena, CA 91109-7009
818-568-6589

Mike Cook
Office of Wastewater
Enforcement & Compliance
U.S. Environmental Protection Agency
401 M Street, SW (WH-546)
Washington, DC 20460
202-382-5850

Glen Daigger
CH2M Hill
P.O. Box 22508
Denver, CO 80222
303-771-0900

Atal Eralp
Office of Wastewater
Enforcement & Compliance
U.S. Environmental Protection Agency
401 M Street, SW (WH-547)
Washington, DC 20460
202-382-7369

Steve Fishel
Division of Water Pollution Control
150 Ninth Avenue North
Nashville, TN 37247
615-741-0633

Bob Freeman
Cobb County Water System
680 South Cobb Drive, SE
Marietta, GA 30060
404-423-1000

Bob Gearheart
Department of Engineering
Humboldt State University
Arcata, CA 95521
707-826-3619

Stephen Hogye
Office of Wastewater
Enforcement & Compliance
U.S. Environmental Protection Agency
401 M Street, SW (WH-547)
Washington, DC 20460
202-382-5841

William Horst
City of Lancaster
120 North Duke Street
Lancaster, PA 17603
717-291-4825

Ancil Jones
Water Management Division
U.S. Environmental Protection Agency
1445 Ross Avenue (6W-MT)
Dallas, TX 75202
214-655-7130

Chris King
Water/Wastewater Division
Crowder College
Route 6
Neosho, MO 64850
417-451-3583

Robert Lee
Office of Wastewater
Enforcement & Compliance
U.S. Environmental Protection Agency
401 M Street, SW (WH-547)
Washington, DC 20460
202-382-7356

Russell Martin
Water Management Division
U.S. Environmental Protection Agency
230 South Dearborn Street
Chicago, IL 60604
312-886-0268

Gary Neuderfer
New York State Department
of Environmental Conservation
6274 East Avon-Lima Road
Avon, NY 14414
716-226-2466

Gordon Nicholson
CH2M Hill
P.O. Box 91500
Bellevue, WA 98009
206-453-5000

Ed O'Brien
Water Quality Program
Washington Department of Ecology
Olympia, WA 98504
206-438-7037

Chuck Pycha
Water Management Division
U.S. Environmental Protection Agency
230 South Dearborn Street
Chicago, IL 60604
312-886-0259

O. Karl Scheible
HydroQual, Inc.
1 Lethbridge Plaza
Mahwah, NJ 07430
201-529-5151

Lori Stone
Engineering-Science, Inc.
10521 Rosehaven Street
Fairfax, VA 22030
703-591-7575

John Sutherland
Division of Environmental Quality
Idaho Department of Health & Welfare
2110 Ironwood Parkway
Coeur d'Alene, ID 83814
208-667-3524

Jim Van Domelen
Department of Environmental Quality
Water Quality - 5th floor
811 Southwest Sixth Street
Portland, OR 97204
503-229-5310

Tom Wooters
Water/Wastewater Division
Crowder College
Route 6
Neosho, MO 64850
417-451-3583

Appendix C—List of Addresses for Regional and State Wastewater Technology, Sludge, and Outreach Coordinators

U.S. EPA REGION	TECHNOLOGY CONTACT	SLUDGE CONTACT	OUTREACH CONTACT
REGION I			
U.S. EPA Water Management Division JFK Federal Building Boston, MA 02203	Charles Conway (617) 565-3517 (FTS) 835-3517	Charles Conway (617) 565-3517 (FTS) 835-3517	Mark Malone (617) 565-3492 (FTS) 835-3492
Connecticut			
Connecticut Department of Environmental Protection 122 Washington Street Hartford, CT 06106	William Hogan (203) 566-2793	Warren Herzig (203) 566-3282	Dennis Greci (203) 566-3282
Maine			
Department of Environmental Protection State House (Station 17) Augusta, ME 04333	Dennis Purington (207) 289-7764	Brian Kavanah (207) 582-8740	William Brown (207) 289-7804

APPENDIX C (Continued)

U.S. EPA REGION	TECHNOLOGY CONTACT	SLUDGE CONTACT	OUTREACH CONTACT
REGION I (Continued)			
Massachusetts			
Division of Water Pollution Control Massachusetts Department of Environmental Quality Engineering One Winter Street Boston, MA 02108	Robert Cady (617) 292-5713	Rick Dunn (617) 556-1130	Howard Bacon (617) 292-5711
New Hampshire			
New Hampshire Department of Environmental Services P.O. Box 95 Concord, NH 03302	John Bush (603) 271-2001	George Berlandi (603) 271-2457	Brad Foster (603) 271-3503
Rhode Island			
Rhode Island Division of Water Resources 291 Promenade Street Providence, RI 02908	Warren Town (401) 277-3961	Chris Campbell (401) 277-3961	David Chopy (401) 277-3961

APPENDIX C (Continued)

U.S. EPA REGION	TECHNOLOGY CONTACT	SLUDGE CONTACT	OUTREACH CONTACT
REGION I (Continued)			
Vermont			
Vermont Agency of Environmental Conservation 103 South Main Street, Bldg. 9 South Waterbury, VT 05676	Marilyn Davies (802) 244-8744	George Desch (802) 244-8744	Jon Jewett (802) 244-8744
REGION II			
U.S. EPA Water Management Division 26 Federal Plaza, Room 837 New York, NY 10278	John Mello (212) 264-5677 (FTS) 264-5677	Aristotle Harris (212) 264-4707 (FTS) 264-4707	Muhammad Hatim (212) 264-8969 (FTS) 264-8969
New Jersey			
New Jersey Department of Environmental Protection P.O. Box CN-029 Trenton, NJ 08625	Robert Kotch (609) 292-6894	Mary Jo Aiello (609) 633-3823	Barbara Hirst (609) 633-1170

APPENDIX C (Continued)

U.S. EPA REGION	TECHNOLOGY CONTACT	SLUDGE CONTACT	OUTREACH CONTACT
REGION II (Continued)			
New York			
Technical Assistance Section New York State Department of Environmental Conservation 50 Wolf Road Albany, NY 12233	Charles Rudick (518) 457-3824	Rick Hammand (518) 457-2051	Diana Perley (518) 457-3810
Puerto Rico			
Puerto Rico Environmental Quality Board Banco Nacional Building 431 Ponce De Leon Blvd. Hato Rey, PR 00913	Baltazar Luna (809) 767-2255	Ava Hernandez (809) 767-2447	Tomas Rivera (809) 767-8073
Virgin Islands			
Natural Resources Management Office 179 Altoona and Welqunst Charlotte Amalie, St. Thomas Virgin Islands 00801	Leonard Reed (809) 774-3320		Francine Lang

APPENDIX C (Continued)

U.S. EPA REGION	TECHNOLOGY CONTACT	SLUDGE CONTACT	OUTREACH CONTACT
REGION III			
U.S. EPA Water Management Division 841 Chestnut Street Philadelphia, PA 19107	Clyde Turner (215) 597-8223 (FTS) 597-8223	Kenneth Pantuck (215) 597-9478 (FTS) 597-9478	Bob Runowksi (215) 597-6526 (FTS) 597-6526
Delaware			
Delaware Department of Natural Resources and Environmental Control Division of Water Resources 89 Kings Highway, Box 1401 Dover, DE 19903	Roy R. Parikh (302) 739-5081	William Razor (302) 739-4781	Roy R. Parikh (302) 739-5081
District of Columbia			
District of Columbia Department of Public Works Water and Sewer Utility Administration 5000 Overlook Avenue, S.W. Washington, DC 20032	Leonard R. Benson (202) 767-7603	Leonard R. Benson (202) 767-7603	
Maryland			
Department of Environment Water Management Administration 2500 Broening Highway Baltimore, MD 21224	John Milnor (301) 631-3726	Doug Proctor (301) 631-3375	Marie Halka (301) 631-3599

APPENDIX C (Continued)

U.S. EPA REGION	TECHNOLOGY CONTACT	SLUDGE CONTACT	OUTREACH CONTACT
REGION III (Continued)			
Pennsylvania			
Pennsylvania Department of Environmental Resources Division of Municipal Facilities and Grants P.O. Box 2063 Harrisburg, PA 17120	Charles Kuder (717) 787-3481	William Pounds (717) 787-7381	Ted Fasting (717) 787-3481
Virginia			
Virginia State Water Control Board Box 11143 Richmond, VA 23230	Walter Gills (804) 367-8860	Cal M. Sawyer (804) 786-1755	Donald Wampler (804) 257-1025
West Virginia			
West Virginia Department of Natural Resources Division of Water Resources 617 Broad Street Charleston, WV 25301	Elbert Morton (304) 348-0633	Clifton Browning (304) 348-2108	Michael Johnson (304) 348-0641

APPENDIX C (Continued)

U.S. EPA REGION	TECHNOLOGY CONTACT	SLUDGE CONTACT	OUTREACH CONTACT
REGION IV			
U.S. EPA Water Management Division 345 Courtland Street, N.E. Atlanta, GA 30365	John Harkins (404) 347-3633 (FTS) 257-3633	Vince Miller (404) 347-3633 (FTS) 257-3633	Roger DeShane (404) 347-3633 (FTS) 257-3633
Alabama			
Alabama Department of Environmental Management 1751 Federal Drive Montgomery, AL 36130	David Hutchinson (205) 271-7761	Cliff Evans (205) 271-7761	Dennis Harrison (205) 271-7801
Florida			
Bureau of Wastewater Management and Grants Florida Department of Environmental Regulation Twin Towers Office Building 2600 Blair Stone Road Tallahassee, FL 32399-2400	Bhupendra Vora (904) 488-8163	J.N. Ramaswamy (904) 488-8163	Jim Battone (904) 488-4524
Georgia			
Environmental Protection Division Georgia Department of Natural Resources Floyd Towers East, Suite 1058 205 Butler Street, S.E. Atlanta, GA 30334	Ernest Earn (404) 656-4769	Mike Creason (404) 656-4887	Ernest Earn (404) 656-4769

APPENDIX C (Continued)

U.S. EPA REGION	TECHNOLOGY CONTACT	SLUDGE CONTACT	OUTREACH CONTACT
REGION IV (Continued)			
Kentucky			
Kentucky Department of Environmental Protection Division of Water 18 Reilly Road Frankfort, KY 40601	Vince Borres (502) 564-3410	Art Curtis (502) 564-3410	Vince Borres (502) 564-3410
Mississippi			
Mississippi Departmennt of Environmental Quality Office of Pollution Control P.O. Box 10385 Jackson, MS 39289-0385	Sitaram Makena (601) 961-5171	Billy Warden (601) 961-5060	Sitaram Makena (601) 961-5171
North Carolina			
Division of Environmental Management North Carolina Department of Natural Resources and Community Development P.O. Box 27687 Raleigh, NC 27611	Allen Wahab (919) 733-6900	Dennis Ramsey/ Allen Wahab (919) 733-6900	Eric Stockton (919) 733-6900

APPENDIX C (Continued)

U.S. EPA REGION	TECHNOLOGY CONTACT	SLUDGE CONTACT	OUTREACH CONTACT
REGION IV (Continued)			
South Carolina			
Domestic Wastewater Division South Carolina Department of Health and Environmental Control 2600 Bull Street Columbia, SC 29201	Sam Grant (803) 734-5279	Mike Montebello (803) 734-5262	Sam Grant (803) 734-5279
Tennessee			
Tennessee Department of Health and Environment Terra Building, 3rd Floor 150 Ninth Avenue North Nashville, TN 37219-5404	Sam Gaddipati (615) 741-0638	Steve Sanford (615) 741-0638	Bill Dobbins (615) 741-0638
REGION V			
U.S. EPA Water Management Division 230 South Dearborn Street Chicago, IL 60604	Charles Pycha (312) 886-0259 (FTS) 886-0259	John O'Grady (312) 353-1938 (FTS) 353-1938	Al Krause (312) 886-0246 (FTS) 886-0246

APPENDIX C (Continued)

U.S. EPA REGION	TECHNOLOGY CONTACT	SLUDGE CONTACT	OUTREACH CONTACT
REGION V (Continued)			
Illinois			
Division of Water Pollution Control Illinois Environmental Protection Agency 2200 Churchill Road Springfield, IL 62706	Ward Akers (217) 782-1696	Al Keller (217) 782-1696	William H. Busch (217) 782-1696
Indiana			
Special Projects Section Water Management Division Indiana Department of Environmental Management 105 South Meridian Street, P.O. Box 6015 Indianapolis, IN 46206-6015	Donald Daily (317) 232-8636	Patrick Carroll (317) 232-8736	Tom Kessling Jeff Fellow (317) 232-8624/8619
Michigan			
Michigan Department of Natural Resources State Office Bldg., Sixth Floor 350 Ottawa St., N.W. Grand Rapids, MI 49503/ Box 30028 Lansing, MI 48909 (Sludge & Outreach)	Ron Woods (616) 456-5071	Al Howard (517) 373-9523	Thomas L. Kamppinen (517) 373-0997

APPENDIX C (Continued)

U.S. EPA REGION	TECHNOLOGY CONTACT	SLUDGE CONTACT	OUTREACH CONTACT
REGION V (Continued)			
Minnesota			
Municipal Section Division of Water Quality Minnesota Pollution Control Agency 520 Lafayette Road St. Paul, MN 55155	Lori Frekot (612) 296-8762	Jorja DuFresue (612) 296-9292	Dave Nelson (612) 296-9274
Ohio			
Division of Environmental & Financial Assistance Ohio Environmental Protection Agency P.O. Box 1049 1800 Water Mark Drive Columbus, OH 43266-10349	Margaret Klepic (614) 644-2828	Stuart M. Blydenburgh (614) 644-2001	Sanat K. Barua (614) 644-2798
Wisconsin			
Municipal Wastewater Section Wisconsin Department of Natural Resources 101 S. Webster Street P.O. Box 7921 Madison, WI 53707	Bob Steindorf (608) 266-0449	Tom Portle (608) 266-8343	Roger Larson (608) 266-2666

APPENDIX C (Continued)

U.S. EPA REGION	TECHNOLOGY CONTACT	SLUDGE CONTACT	OUTREACH CONTACT
REGION VI			
U.S. EPA Water Management Division Allied Bank Tower at Fountain Place 1445 Ross Avenue Dallas, TX 75202	Ancil Jones (214) 655-7130 (FTS) 255-7130	Ancil Jones (214) 655-7130 (FTS) 255-7130	Gene Wossum (214) 655-7130 (FTS) 255-7130
Arkansas			
Arkansas Department of Pollution Control and Ecology, P.O. Box 9583 8001 National Drive Little Rock, AR 72209	Martin Roy (501) 562-8910	Mike Hood (501) 562-8910	Daniel J. Clanton (501) 562-8910
Louisiana			
Louisiana Department of Environmental Quality P.O. Box 82263 Baton Rouge, LA 70884-2263	Robert Crawford (504) 765-0810	Ken Fledderman (504) 342-0067	Thomas Griggs (504) 342-0067
New Mexico			
New Mexico Environmental Improvement Agency Water Quality Section Harold Runnels Building 1190 St. Francis Drive P.O. Box 968 Santa Fe, NM 87501-0968	Arun Dhawan (505) 827-2809	Arun Dhawan (505) 827-2809	David Hanna (505) 827-2812

APPENDIX C (Continued)

U.S. EPA REGION	TECHNOLOGY CONTACT	SLUDGE CONTACT	OUTREACH CONTACT
REGION VI (Continued)			
Oklahoma			
Engineering Division Oklahoma State Department of Health 3400 North Eastern Avenue P.O. Box 53551 Oklahoma City, OK 73152	H.J. Thung (405) 271-7346	Danny Hodges (405) 271-7362	Jon L. Craig (405) 271-5205
Texas			
Texas Water Development Board 4615 Hawkhaven Lane Austin, TX 78727	Milton Rose (512) 463-8513	Milton Rose (512) 463-8513	Michael McDevitt Milton Rose (409) 463-8503/8513
REGION VII			
U.S. EPA Water Management Division 726 Minnesota Avenue Kansas City, KS 66101	Rao Surampalli (913) 551-7453 (FTS) 276-7453	Rao Surampalli (913) 551-7453 (FTS) 276-7453	Kelly Beard Tittone (913) 551-7217 (FTS) 276-7217

APPENDIX C (Continued)

U.S. EPA REGION	TECHNOLOGY CONTACT	SLUDGE CONTACT	OUTREACH CONTACT
REGION VII (Continued)			
Iowa			
Program Operations Division Iowa Department of Natural Resources Environmental Protection Division Henry A. Wallace Building 900 East Grand Des Moines, IA 50319	Terry Kirschenman (515) 281-8885	Darrell McAllister (515) 281-8869	Darrell McAllister (515) 281-8869
Kansas			
Municipal Programs Section Division of Environment Kansas Department of Health and Environment Forbes Field, Bldg. 740 Topeka, KS 66620-0110	Rodney Geisler (913) 296-5527	Rodney Geisler (913) 296-5527	Jack Alexander (913) 296-5513
Missouri			
Water Pollution Control Program Division of Environmental Quality Missouri Department of Natural Resources P.O. Box 176 Jefferson City, MO 65102	Randy Clarkson (314) 751-6620	Ken Arnold (314) 751-6624	Gerry Fields (314) 751-7537

APPENDIX C (Continued)

U.S. EPA REGION	TECHNOLOGY CONTACT	SLUDGE CONTACT	OUTREACH CONTACT
REGION VII (Continued)			
Nebraska			
Construction Grants Branch Water Quality Section Nebraska Department of Environmental Control P.O. Box 98922 Statehouse Station Lincoln, NE 68509-8922	Mahmood Arbab (402) 471-4236	Rick Bay (402) 471-2186	Gautan Bhadbhade (402) 471-4207
REGION VIII			
U.S. EPA Water Management Division 999 - 18th Street, Suite 500 Denver, CO 80202-2405	Jim Brooks (303) 293-1549 (FTS) 330-1549	Bob Brobst (303) 293-1627 (FTS) 330-1627	Pauline Afshar (303) 293-1176 (FTS) 330-1176
Colorado			
Water Quality Control Division Colorado Department of Health 4210 E. 11th Avenue Denver, CO 80220	Derald Lang (303) 331-4564	Phil Hegeman (303) 331-4564	J. David Holm (303) 331-4830

APPENDIX C (Continued)

U.S. EPA REGION	TECHNOLOGY CONTACT	SLUDGE CONTACT	OUTREACH CONTACT
REGION VIII (Continued)			
Montana			
Water Quality Bureau Montana Department of Health and Environmental Sciences Cogswell Building Helena, MT 59620-0522	Scott Anderson (406) 444-2406	Scott Anderson (406) 444-2406	Scott Anderson (406) 444-2406
North Dakota			
Division of Water Supply and Pollution Control North Dakota Department of Health 1200 Missouri Avenue, P.O. Box 5520 Bismark, ND 58502-5520	Jeff Hauge (701) 221-5220	Jeff Hauge Sheila McClenathan (701) 221-5220/5227	Jeff Hauge (701) 221-5220
South Dakota			
South Dakota Department of Water and Natural Resources Joe Foss Building, 523 East Capitol Pierre, SD 57501-3181	Curt Struck (605) 773-4216	Bill Geyer (605) 773-3351	Curt Struck (605) 773-4216

APPENDIX C (Continued)

U.S. EPA REGION	TECHNOLOGY CONTACT	SLUDGE CONTACT	OUTREACH CONTACT
REGION VIII (Continued)			
Utah			
Utah Bureau of Water Pollution Control P.O. Box 16690 Salt Lake City, UT 84116-0690	Kiran L. Bhayani (801) 533-6048	Kiran L. Bhayani Paul Krauth (801) 533-6048/6146	Walt Baker (801) 538-6146
Wyoming			
Water Quality Division Wyoming Department of Environmental Quality Herschler Bldg., 4 West 122 W. 25th Street Cheyenne, WY 82002	Mike Hackett (307) 777-7781	E.J. Fanning John Wagner (307) 777-7781/7081	Mike Hackett (307) 777-7781
REGION IX			
U.S. EPA Water Management Division 75 Hawthorne Street San Francisco, CA 94105	Tamara Rose (415) 744-1945 (FTS) 484-1945	Lauren Fondahl (415) 744-1909 (FTS) 484-1909	Carla Moore (415) 744-1935 (FTS) 484-1935
Arizona			
Arizona Department of Environmental Quality 2005 North Central Avenue Phoenix, AZ 85004	Ron Frey (602) 257-2220	Barry Abbott (602) 257-2176	Bill Shafer (602) 257-2220

APPENDIX C (Continued)

U.S. EPA REGION	TECHNOLOGY CONTACT	SLUDGE CONTACT	OUTREACH CONTACT
REGION IX (Continued)			
California			
State Water Resources Control Board Division of Clean Water Programs P.O. Box 944212 Sacramento, CA 94224-2120	Gordon Innes (916) 327-5567	Archie Mathews (916) 322-4507	Jim Putman (916) 739-2725
Hawaii			
Construction Grants Program Hawaii State Department of Health 5 Water Front Plaza, Suite 250 500 Ala Moana Blvd. Honolulu, HI 96813	Dennis Tulang (808) 543-8288	Dennis Tulang (808) 543-8288	Dennis Tulang (808) 543-8288
Nevada			
Nevada Department of Environmental Protection - Construction Grants Capitol Complex 123 W. Nye Lane Carson City, NV 89710	James Williams (702) 687-5870	James Williams (702) 687-5870	James Williams (702) 687-4670

APPENDIX C (Continued)

U.S. EPA REGION	TECHNOLOGY CONTACT	SLUDGE CONTACT	OUTREACH CONTACT
REGION X			
U.S. EPA Water Management Division 1200 Sixth Avenue Seattle, WA 98101	Bryan Yim (206) 553-8575 (FTS) 399-8575	Dick Hetherington (206) 553-1941 (FTS) 399-1941	Bryan Yim (206) 553-8575 (FTS) 399-8575
Alaska			
Alaska Department of Environmental Conservation Division of Water Programs P.O. Box O Juneau, AK 99811-1800	Richard Marcum (907) 465-2610	Stan Hungerford (907) 465-2610	Linda Taylor (907) 465-2620
Idaho			
Idaho Division of Environmental Quality 1420 N. Hilton Street Boise, ID 83706-1260	Bob Braun (208) 334-5860	Susan Martin (208) 334-5855	Al Stanford (208) 334-5855
Oregon			
Oregon Department of Environmental Quality 811 SW 6th Avenue Portland, OR 97204-1390	Francis Dzata (503) 229-6750	Richard Nichols (503) 229-5324	Martin Loring (503) 229-5415

APPENDIX C (Continued)

U.S. EPA REGION	TECHNOLOGY CONTACT	SLUDGE CONTACT	OUTREACH CONTACT
REGION X (Continued)			
Washington			
Department of Ecology 4500 Third Avenue (PV-11) Olympia, WA 98504	Al Newman (206) 459-6088	Jim Knudson (206) 459-6597	Carl Jones (206) 438-7044

Appendix D—Summary of Innovative and Alternative Technology Projects by State

EPA REGION	STATE	Aeration						Clarification					Collection	
		Counter Current Aeration	Draft Tube Aeration	Fine Bubble Diffusers	Aero-mod System	Intermittent Cycle Extended Aeration	Other Aeration	Flocculating Clarifiers	Integral Clarifiers	Intrachannel Clarifiers	Swirl Concentrators	Other Clarification	Small Diameter Gravity Sewers	Other Collection Systems
I	Connecticut			1										
	Maine		1				1				1	1		
	Massachusetts			1										
	New Hampshire													
	Rhode Island		1											
	Vermont													
II	New Jersey											1		
	New York		3						2				1	1
	Puerto Rico													
	Virgin Islands													
III	Delaware		1							1				
	Washington DC										1			
	Maryland									1				
	Pennsylvania	1	2									1		2
	Virginia	1	1						1	2			1	
	West Virginia		2				1		1	3		1		
IV	Alabama	5	4	2						4				
	Florida	1												
	Georgia	1												
	Kentucky	1								7			1	
	Mississippi									1				
	North Carolina	5	2		1		1							
	South Carolina	1								1				
	Tennessee	7				4				2	1			1
V	Illinois			1						2	1			
	Indiana			2						1	3			
	Michigan						1					1		
	Minnesota						1			1				
	Ohio			2	1					3	1	1	4	1
	Wisconsin			2				2						
VI	Arkansas						1			1				
	Lousiana	1					1			4				
	New Mexico		1											
	Oklahoma					2	1			1				
	Texas						2			2				
VII	Iowa						2			1				
	Kansas		1							1				
	Missouri				3					4				
	Nebraska													
VIII	Colorado													
	Montana							1					1	
	North Dakota													
	South Dakota									2				
	Utah							1						
	Wyoming													
IX	Arizona											1		
	California											2		
	Guam													
	Trust Territories													
	Hawaii													
	Nevada													
	N. Marianas Islands													
X	Alaska													
	Idaho		1							1				
	Oregon											1		
	Washington													
TOTAL		24	20	11	5	6	12	4	4	46	8	10	8	5

SUMMARY OF INNOVATIVE TECHNOLOGY PROJECTS (cont'd)

EPA REGION	STATE	Disinfection		Disposal of Effluent	Energy Conservation and Recovery		Filtration			Lagoons			
		Ultraviolet Disinfection	Other Disinfection	Other Disposal of Effluent	Solar Heating	Other Energy Conservation and Recovery	Biological Aerated Filters	Microscreens	Other Filtration	Aquaculture	Hydrograph Controlled Release Lagoons	Single Cell Lagoon with Sand Filter	Other Lagoons
I	Connecticut	2											
	Maine	1	2		1				1		3		
	Massachusetts	1	1		1								
	New Hampshire	1											
	Rhode Island				1								
	Vermont	1			1								1
II	New Jersey												
	New York	4				1							
	Puerto Rico												
	Virgin Islands												
III	Delaware												
	Washington DC												
	Maryland	3	1										
	Pennsylvania								2				
	Virginia	1	1					1	2	1			
	West Virginia												
IV	Alabama	1					1				5		
	Florida			2									
	Georgia				1	1							
	Kentucky								1		2		
	Mississippi							1			8		
	North Carolina						1						
	South Carolina				1		1				1		
	Tennessee								1		2		
V	Illinois								2			10	2
	Indiana												
	Michigan								1				
	Minnesota	5	1		1	4			1				1
	Ohio	3	1		1	1			1				
	Wisconsin	1				3			1		1		
VI	Arkansas	1								2	1		
	Lousiana	4								1	1		2
	New Mexico			1									
	Oklahoma	4				1							1
	Texas			1						2			
VII	Iowa	1											
	Kansas	3									1		1
	Missouri	1	1			1			2				
	Nebraska					1		1					
VIII	Colorado	1						1					
	Montana	4	1										
	North Dakota												
	South Dakota												1
	Utah												
	Wyoming	4			1				2				1
IX	Arizona	1	1		1	1							1
	California		1	1		3			3				1
	Guam												
	Trust Territories												
	Hawaii												
	Nevada					1		1					
	N. Marianas Islands												
X	Alaska												1
	Idaho							1					1
	Oregon					1							
	Washington	1				1							
TOTAL		49	11	5	10	20	3	6	20	6	25	10	14

SUMMARY OF INNOVATIVE TECHNOLOGY PROJECTS (cont'd)

EPA REGION	STATE	Land Application of Effluent: Overland Flow	Land Application of Effluent: Other Land Application of Effluent	Nitrification: Other Nitrification	Nutrient Removal: Anoxic/oxic system (A/O)	Nutrient Removal: Phostrip	Nutrient Removal: Sequencing Batch Reactor (SBR)	Nutrient Removal: Other Nutrient Removal	Oxidation Ditch: Barrier Wall Oxidation Ditch	Oxidation Ditch: Other Oxidation Ditch	Fixed Growth: Rotating Biological Contactors (RBC's)	Fixed Growth: Trickling Filter/Solids Contact	Fixed Growth: Other Fixed Growth
I	Connecticut												
	Maine		1										
	Massachusetts			2		1					1		
	New Hampshire												
	Rhode Island												
	Vermont		1										
II	New Jersey									1	1		
	New York			1		2			3	3		1	
	Puerto Rico												
	Virgin Islands												
III	Delaware								1				
	Washington DC												
	Maryland				1		1	2					
	Pennsylvania		1		1		3			2			
	Virginia		1	1				1		6			
	West Virginia									1			1
IV	Alabama									5			
	Florida		1		1			1					
	Georgia		1										
	Kentucky	2									1		
	Mississippi		1										
	North Carolina									2			
	South Carolina		1				2			1			
	Tennessee	1					1			3			
V	Illinois	1						1		1		2	
	Indiana			1									
	Michigan				1					2		1	
	Minnesota		1	1		1		2		1			
	Ohio							1		1		1	
	Wisconsin												
VI	Arkansas	3			1			1				1	
	Lousiana	1	1							5			
	New Mexico											1	
	Oklahoma	1					6			2	1	1	
	Texas	3								1	1	2	
VII	Iowa						3			1			
	Kansas									1			
	Missouri						3						
	Nebraska												
VIII	Colorado			2			1						
	Montana		1					2					
	North Dakota		1										
	South Dakota		1				1			1			
	Utah												
	Wyoming												
IX	Arizona							1					
	California			1				1					
	Guam												
	Trust Territories												
	Hawaii												
	Nevada					1							
	N. Marianas Islands												
X	Alaska						1						
	Idaho									1		1	
	Oregon		1		1							1	
	Washington												
TOTAL		12	14	9	6	5	22	13	4	41	5	12	1

SUMMARY OF INNOVATIVE TECHNOLOGY PROJECTS (cont'd)

EPA REGION	STATE	Sludge Technologies						Onsite Technologies	Miscellaneous		Suspended Growth	
		Carver-Greenfield	Composting	Digestion	Incineration	Vacuum Assisted Sludge Drying Beds	Other Sludge Technologies	Other Onsite Technologies	Enclosed Impellor Screw Pump	Other Miscellaneous	Powdered activated Carbon/Regeneration	Other Suspended Growth
I	Connecticut		2									
	Maine		11							2		
	Massachusetts								1	1		
	New Hampshire											
	Rhode Island											
	Vermont											
II	New Jersey	1	1				4			1		
	New York		1		1					1		
	Puerto Rico											
	Virgin Islands											
III	Delaware		3									
	Washington DC									1		
	Maryland		6				1			1		
	Pennsylvania					1	1			1		
	Virginia					1				1		
	West Virginia			1								
IV	Alabama		1				2			1		
	Florida		14									
	Georgia		2									
	Kentucky		2				1					
	Mississippi											
	North Carolina			1	1	1						
	South Carolina		2				1					
	Tennessee									2		
V	Illinois					1					1	
	Indiana		2			2						
	Michigan						1			1	1	
	Minnesota			2		1		1				
	Ohio		1			1					2	
	Wisconsin							1				
VI	Arkansas						1	1		1		
	Lousiana					1						1
	New Mexico						2					
	Oklahoma			1		1						
	Texas					1	3			4		
VII	Iowa		1	1								
	Kansas		2				2		1			
	Missouri				2				2	1		
	Nebraska					1	1					
VIII	Colorado		4			1						
	Montana			1		1						
	North Dakota											
	South Dakota											
	Utah											
	Wyoming											
IX	Arizona						1					
	California	2	5			1	1			2		
	Guam											
	Trust Territories											
	Hawaii		1							1		
	Nevada											
	N. Marianas Islands											
X	Alaska		1		1		2			1		
	Idaho			1						1		
	Oregon						1					
	Washington											
TOTAL		3	62	8	5	15	25	3	4	24	4	1

SUMMARY OF ALTERNATIVE TECHNOLOGY PROJECTS

		ONSITE TREATMENT						SOLIDS TREATMENT						
EPA REGION	STATE	Septic Tank/Soil Absorption System (Single Family)	Mound System	Evapotranspiration Bed	Aerobic Unit	Sand Filter	Other Onsite Treatment	Septage Treatment and Disposal	Land Spreading of POTW Sludge	Composting	Preapplication Treatment	90% Methane Recovery from Anaerobic Digestion	Self-sustaining Incineration (Heat Recovery and Utilization)	Other Sludge Treatment or Disposal
I	Connecticut					1		7		1		4	1	
	Maine	4				5				6				
	Massachusetts					1		19	1	5		2	2	
	New Hampshire					4		7		1				
	Rhode Island							2		1			1	
	Vermont					2			12					
II	New Jersey							11	1	5	3			1
	New York	1	2			12	1	4	2	2		16	1	1
	Puerto Rico									1				
	Virgin Islands													
III	Delaware								2	2	1			
	Washington DC									1				
	Maryland	3	2			2	1	3	4	5		2		
	Pennsylvania	3	1		1	1			6	3		3		2
	Virginia							3	9	3	1	5	2	3
	West Virginia	1					1					2		
IV	Alabama								2	1		3		
	Florida									2	1	3		
	Georgia								4			4	1	
	Kentucky	1				2			13			2		
	Mississippi								3					1
	North Carolina								4			6		1
	South Carolina								5			1		
	Tennessee	2							5	1				1
V	Illinois	4	1			15	1	2	44		17	13		5
	Indiana	1	2					1	20		3	5	1	
	Michigan	2						1	16		3	4	1	
	Minnesota	10	9			2	1		24	2	3	6		
	Ohio	1	1			1		3	35	4	9	6	1	
	Wisconsin		3			1			13			3		1
VI	Arkansas	1						1	2			1		
	Lousiana				1				1					
	New Mexico										1	1		1
	Oklahoma						1		10	1	2		1	
	Texas	2			1				29	1	8	7		1
VII	Iowa					1			24		2	6		
	Kansas								20	1		5	2	2
	Missouri		1			1		1	34			1		8
	Nebraska								5	2	3	3		2
VIII	Colorado								2			1		
	Montana								9			2		1
	North Dakota		4											
	South Dakota								11		1	4		
	Utah								1	1		1		
	Wyoming								2			1		
IX	Arizona								1			3		1
	California		1					4	1	1	2	5	1	2
	Guam													
	Trust Territories						1							
	Hawaii													
	Nevada											3		
	N. Marianas Islands													
X	Alaska								1	1		1	1	1
	Idaho					2	1		6			3		
	Oregon			2		3			4	1	3	5		2
	Washington		1						1		1	2		
TOTAL		36	28	2	3	56	8	69	389	55	64	145	16	37

SUMMARY OF ALTERNATIVE TECHNOLOGY PROJECTS (cont'd)

		TREATMENT/DISCHARGE SYSTEMS										COLLECTION SYSTEMS			
EPA REGION	STATE	Overland Flow	Rapid Infiltration Land Treatment Systems	Slow Rate Treatment Systems	Other Land Treatment Systems	Septic Tank/Soil Absorption System (Multiple Families)	Preapplication Treatment or Storage	Total Containment Ponds	Aquaculture/Wetlands/ Marsh Systems	Aquifer Recharge	Direct Reuse	Pressure Sewers, Septic Tank Effluent Pump (STEP)	Small Diameter Gravity Sewers	Pressure Sewers, Grinder Pump (GP)	Vacuum Sewers
I	Connecticut		1			2									
	Maine			1		7						1		1	
	Massachusetts		2	1									1		
	New Hampshire	1			1	3								1	
	Rhode Island														
	Vermont			1		1							1	3	
II	New Jersey		1			1	1							2	2
	New York	2	3			2						3	16	16	3
	Puerto Rico														
	Virgin Islands														
III	Delaware		1			2							1	2	
	Washington DC														
	Maryland	2	1	5	2		4	1	2			2	3	27	6
	Pennsylvania		1	5	1	2	2		2			6	13	15	
	Virginia	3	1	1							5	3	2	2	2
	West Virginia											6	5	10	11
IV	Alabama			2								1	3	2	
	Florida	1	2	20						1	3				2
	Georgia	2	1	18								2	1		
	Kentucky	1		2		2			2			2	5	4	2
	Mississippi	11		2								1	1	3	
	North Carolina			21								1	1	2	
	South Carolina		1	9									2		1
	Tennessee	2		6			4					5	10	6	3
V	Illinois	4	1	3	1	1	3				5	4	21	3	
	Indiana			1		2						3	14	4	5
	Michigan	4	4	14		7	13		1				2	3	
	Minnesota		1	15		9	15					6	7	4	
	Ohio			1			1					3	2	6	
	Wisconsin		17	1	1		9					2	4	2	
VI	Arkansas			5	1		2					1	2	10	
	Lousiana	6		2			2		1			1	1	1	
	New Mexico			6			5								
	Oklahoma			31			16	29							
	Texas	1	1	11	2		10	1			4	3	1	6	
VII	Iowa			2			3		3			2	1	3	
	Kansas		1	16			9	27	1						
	Missouri	14		25					2			6	15	20	
	Nebraska		2	5				32	1						
VIII	Colorado			2	1		1			1				1	
	Montana		3	8				5							
	North Dakota			6				17				3	14	2	
	South Dakota		8	1			3	7	5		1	1	2	1	
	Utah			3			2								
	Wyoming		2	2				3							
IX	Arizona	1	1	12			1	1	4		1	1	1		
	California	2	14	20		3	25	1	3		2	7	4		1
	Guam		1			1									
	Trust Territories					5							2		
	Hawaii			2				1							
	Nevada		5	6			5	4	1						
	N. Marianas Islands														
X	Alaska					2	1					1	1		2
	Idaho		4	9	1	1	10	1	1			2	2		
	Oregon		1	6	1		9	1	2			5	4		1
	Washington			3	2	1	4	2	1			3	2	1	
TOTAL		57	81	312	14	54	160	133	32	2	21	87	167	163	41

Appendix E—Current Status of Modification/Replacement (M/R) Grant Candidates by State

STATE	COMMUNITY	TECHNOLOGY	M/R GRANT AWARDED	GRANT IN REVIEW	POTEN-TIAL M/R PROJECT	SUB-JECT OF LITIGA-TION
Alabama	Littleville	Intrachannel clarifiers			X	
	Leighton	Intrachannel clarifiers			X	
	Phil Campbell	Intrachannel clarifiers			X	
Arizona	Flagstaff	Combined chlorination/clarification	X			
Arkansas	Paragould	Aquaculture	X			
California	City of Los Angeles	Total energy recovery system		X		
	Fallen Leaf Lake	Vacuum collection system	X			
	Gustine	Aquaculture		X		
	Hayward	Fluidized-bed reactors		X		
	Manila	Community leach field		X		
	Nevada City	Vacuum-assisted sludge drying beds	X			
	Reedley	Rapid infiltration		X		
	San Lorenzo	Pressure leach field for effluent disposal	X			
	Ventura Nyeland Acres	Septic tank effluent pump collection system controllers and pumps		X		
	West Point	Community leach field		X		
Colorado	Idaho Springs	UV disinfection			X	
	Longmont	Breakpoint chlorination			X	
Florida	Ft. Lauderdale	Mechanical composting		X		
Idaho	Grangeville	Draft tube aeration	X			

APPENDIX E (Continued)

CURRENT STATUS OF MODIFICATION/REPLACEMENT (M/R) GRANT CANDIDATES BY STATE

STATE	COMMUNITY	TECHNOLOGY	M/R GRANT AWARDED	GRANT IN REVIEW	POTENTIAL M/R PROJECT	SUBJECT OF LITIGATION
Illinois	Hanover	Sand filters	X			
Indiana	Portage	Vacuum-assisted sludge drying beds	X			
Kansas	Bonner Springs	Intrachannel clarifiers		X		
	Dodge City	Odor control	X			
Kentucky	Berea	Intrachannel clarifier	X			
	Elkton	Side channel clarifier	X			
Maine	Presque Isle	UV disinfection	X			
	Sabattus	UV disinfection	X			
Massachusetts	Fall River	Self-sustaining incineration	X			
	Southbridge	Sludge composting			X	
	Wayland	Septage treatment	X			
	Westboro	Sludge composting			X	
	Williamstown	Grinder pumps/pressure sewers	X			
Michigan	Ionia	Rotating biological contactors		X		

APPENDIX E (Continued)

CURRENT STATUS OF MODIFICATION/REPLACEMENT (M/R) GRANT CANDIDATES BY STATE

STATE	COMMUNITY	TECHNOLOGY	M/R GRANT AWARDED	GRANT IN REVIEW	POTENTIAL M/R PROJECT	SUBJECT OF LITIGATION
Minnesota	Moorehead	Active ozone disinfection	X			
	Northfield	UV disinfection	X			
	North Koochiching	UV disinfection	X			
	Pine River	Sludge composting and rotatating biological contactors			X	
	Rochester	Biological phosphorus removal			X	
Mississippi	Newton	Overland flow			X	
Missouri	Excelsior Springs	Overland flow	X			
	Gallatin	Intrachannel clarifiers		X		
	Little Blue Valley	Intrachannel clarifiers		X		
	Purdy	Storage lagoon/land application			X	
Montana	Bozeman	Rapid infiltration	X			
Nebraska	Scotts Bluff	Microscreen ponds	X			
Nevada	Henderson	Rapid infiltration basins			X	
	Incline Village	Wetlands	X			
New Mexico	Santa Fe	Draft tube aerators	X			
New York	Lawrence	Community mound systems	X			
	Peru	Rapid infiltration basin			X	
	Plattsburgh	In-vessel composting	X			

APPENDIX E (Continued)

CURRENT STATUS OF MODIFICATION/REPLACEMENT (M/R) GRANT CANDIDATES BY STATE

STATE	COMMUNITY	TECHNOLOGY	M/R GRANT AWARDED	GRANT IN REVIEW	POTEN-TIAL M/R PROJECT	SUB-JECT OF LITIGA-TION
North Carolina	Washington	Draft tube aerators	X			
	Starr	Draft tube aerators	X			
	Burlington	Powdered activated carbon treatment	X			
	Greensboro	Starved air incinerator	X			
	Greenville	Counter current aeration		X		
	Henderson	Dual digestion	X			
	Pilot Mountain	Jet aeration oxidation ditches		X		
North Dakota	Antler	Community mound systems	X			
	Buchanan	Community mound systems		X		
	Churchs Ferry	Community mound systems	X			
	Clifford	Community mound systems	X			
Ohio	Akron	In-vessel composting	X			
	Bedford Heights	Powdered activated carbon treatment (PACT)	X			
	Clark Co.	Rotating biological contactors		X		
	Clyde	Intrachannel clarifiers	X			
	Ironton	UV disinfection	X			
	Lake County	Composting	X			
	North Olmstead	Powdered activated carbon treatment	X			
	New Carlisle	Rotating biological contactors	X			
	Lower East Fork	Rotating biological contactors	X			
	Waynesburg	Biodrum/UV		X		
Oregon	South Point	Rotating biological contactors	X			
	Cove Orchard	Community leach field	X			
	Dexter	Recirculating sand filter	X			
	Eugene	Spray irrigation		X		
Rhode Island	Cranston	Draft tube aerators	X			

APPENDIX E (Continued)

CURRENT STATUS OF MODIFICATION/REPLACEMENT (M/R) GRANT CANDIDATES BY STATE

STATE	COMMUNITY	TECHNOLOGY	M/R GRANT AWARDED	GRANT IN REVIEW	POTENTIAL M/R PROJECT	SUBJECT OF LITIGATION
South Carolina	Durban Creek	Intrachannel clarifier		X		
South Dakota	Mina Lake	Community mound systems		X		
	White Rivers	Total containment lagoons	X			
	Pollock	Rapid infiltration system			X	
Tennessee	Claiborne Co.	Counter current aeration		X		
	Memphis	Biofilters	X			
Texas	El Paso	Draft tube aerators	X			
	Levelland	Aeration/oxidation ponds	X			
Virginia	Buena Vista	Vacuum-assisted sludge drying beds		X		
Washington	Black Diamond	Wetlands	X			
	Elbe	Community mound systems	X			
West Virginia	Crab Orchard-MacArthur	Draft tube aerators				X
Wisconsin	Cambellsport	Rapid infiltration	X			
	Hayward	Rapid infiltration			X	
	Wittenberg	Rapid infiltration	X			
	Whitewater	Rotating biological contactors	X			